REPRESENTATION THEOREMS ON BANACH FUNCTION SPACES[1]

by Neil E. Gretsky
University of California, Riverside

CHAPTER I: INTRODUCTION

1. The Problem. The Banach function space $L_\rho(\Omega,\Sigma,\mu)$ is a Banach space of (equivalence classes of) measurable functions on the measure space (Ω,Σ,μ) with ρ as a (function) norm. Banach function spaces, sometimes called Riesz spaces or Köthe-Toeplitz spaces have been studied in [16, 34, 14, 15, 7, 8, 18, 19, 21, 22, 23] with the most exhaustive work being in [20]. They include many of the well-known function spaces such as the Lebesgue spaces and the Orlicz spaces and are well suited for settings in analysis for work in many areas, e.g. integral equations and stochastic transformations.

The problems considered in this paper are the obtaining of integral representations of members of $\mathcal{B}(\mathfrak{X},L_\rho)$ and of $\mathcal{B}(L_\rho,\mathfrak{X})$, where $\mathfrak{X}$ is an arbitrary Banach space and $\mathcal{B}(\mathfrak{Y},Z)$ denotes the spaces of continuous linear operators from $\mathfrak{Y}$ to Z . In Chapter II, under the hypothesis that the function norm satisfies a mild type of averaging condition, a characterization of members of the space $\mathcal{B}(\mathfrak{X},L_\rho)$ is obtained as an integral representation in terms of $\mathfrak{X}^*$ -valued additive set functions of ρ-bounded variation. The result here generalizes the fundamental representation theorems of Dunford and Pettis [5] and their later generalizations (cf [6] , p. 498). In the second section of Chapter II, using the same

____

[1]This paper constitutes the major portion of the author's doctoral dissertation written at Carnegie Institute of Technology under the guidance of Professor M. M. Rao. The author acknowledges a deep debt of gratitude to Dr. Rao.
Received by the editor 10-25-67 and, in revised form, 7-19-68.

averaging condition, we have characterized $\mathfrak{B}(M_\rho, \mathfrak{X})$, where M_ρ is the closed subspace of L_ρ determined by the simple functions of L_ρ . Here the representations are in terms of integrals of scalar functions relative to additive vector-valued set functions of ρ'-bounded variation (where ρ' is the norm dual to ρ) , using the integration theory of [2].

The problem of characterizing $\mathfrak{B}(L_\rho, \mathfrak{X})$ in general appears to be intractable by the present methods. However, if $\mathfrak{X}$ is the scalars, then $\mathfrak{B}(L_\rho, \mathfrak{X}) = L_\rho^*$ admits a very detailed analysis and is considered in Chapter III. There are two characterizations given: one assumes the use of the averaging condition mentioned above, the other does not. Both, however, use a further hypothesis on the structure of the quotient space L_ρ/M_ρ . It is shown that any continuous linear functional on L_ρ can be decomposed into two parts, one of which annihilates M_ρ and the other of which corresponds to a functional on M_ρ . (These are classically termed singular and absolutely continuous, respectively.) The represent- ations of these functionals are then obtained with methods which generalize those used in Orlicz spaces. As a by-product a general representation of functionals on M_ρ has been obtained with essen- tially no restrictions on ρ .

2. Preliminaries. A brief outline of Banach function spaces and some desired results will be presented here. Details of the material below up to definition 13 may be found in [19, 20, 21, 22].

Let (Ω, Σ, μ) be a σ-finite measure space, and let M^+ be the collection of all non-negative measurable functions on Ω equipped with the usual pointwise (a.e.) order. (As usual, functions differing only on μ-null sets will be identified so that the elements of M^+ are in reality equivalence classes of functions.)

DEFINITION 1. A mapping ρ on M^+ to the extended reals is called a _function norm_ if ρ satisfies the following conditions for all f and g in M^+ :

    i) $\rho(f) \geq 0$ and $\rho(f) = 0$ if and only if $f = 0$

    ii) $\rho(\alpha f) = \alpha\rho(f)$ for $\alpha \geq 0$

    iii) $\rho(f + g) \leq \rho(f) + \rho(g)$

    iv) $f \leq g$ in M^+ implies $\rho(f) \leq \rho(g)$.

The definition of ρ is extended to M , the collection of all complex valued functions on Ω , by defining $\rho(f) = \rho(|f|)$ for $f \in M$. In order to avoid pathological cases, only non-trivial ρ will be considered, i.e. we require that there exists $f \in M$ such that $0 < \rho(f) < \infty$.

DEFINITION 2. i) A function norm ρ has the _weak Fatou property_ (WFP) if $f_n \uparrow f$ and $\sup_n \rho(f_n) < \infty$ imply $\rho(f) < \infty$ where $f \in M^+$ and $f_n \in M^+$, $n = 1,2,\dots$.

    ii) A function norm ρ has the _strong Fatou property_ (SFP) if $f_n \uparrow f$ implies $\rho(f_n) \uparrow \rho(f)$ where $f \in M^+$ and $f_n \in M^+$, $n = 1,2,\dots$. Note that SFP implies WFP, but the converse need not be true.

DEFINITION 3. Let $L_\rho(\Omega,\Sigma,\mu) = \{f \in M \mid \rho(f) < \infty\}$. It is clear that L_ρ is a normed linear space and, in fact,

is a (complex) AB-lattice. In general, however, it is not complete
without strengthening the hypotheses on ρ . Examples may be
found in [20]. A sufficient condition turns out to be the weak
Fatou property.

  In what follows, it is always assumed without further mention
that ρ has the weak Fatou property.

  DEFINITION 4. A sequence of sets $\{\Omega_n\}$ such that $\Omega_n \in \Sigma$,
$\Omega_n \uparrow \Omega$, $\mu(\Omega_n) < \infty$, and $\rho(\chi_{\Omega_n}) < \infty$ is called __admissible__ with respect
to ρ .

  In order to guarantee the existence of admissible sequences
the underlying measure space may have to be adjusted as follows.

  DEFINITION 5. A set $E \subseteq \Omega$ is called __unfriendly__ (relative
to ρ) if for every measurable $B \subseteq E$, $\mu(B) > 0$ implies $\rho(\chi_B) = \infty$.

  It can be shown that there is a measurable $E_o \subseteq \Omega$ which is
maximal with respect to the property of being unfriendly. This
set is removed from Ω and the new σ-finite measure space will
be relabeled as (Ω,Σ,μ). Note that the removal of E_o does not
really restrict the generality since, if $f \in L_\rho$ and E is unfriendly, then $f(\omega) = 0$ for almost all $\omega \in E$. In the "new"
(Ω,Σ,μ) there are always admissible sequences.

  THEOREM 6. __Let__ L_ρ __be a function space on a__ σ-__finite measure__
__space__ (Ω,Σ,μ) __with no unfriendly sets as above. Then__
a) __There is a constant__ γ , $0 < \gamma \le 1$, __such that if__ $\{f_n\}$ __is__
__any sequence in__ M^+ , $f \in M^+$, __and__ $f_n \uparrow f$, __then__ $\gamma\rho(f) \le \lim_n \rho(f_n)$;
b) L_ρ __is complete (and is called a Banach function space).__

<u>The constant</u> $\gamma = 1$ <u>if and only if</u> ρ <u>has SFP</u>.

DEFINITION 7. <u>The dual norm</u> ρ' of a function norm ρ is defined by $\rho'(f) = \sup \{\int_\Omega |fg| d\mu \mid \rho(g) \le 1\}$. Then the <u>dual space</u> is defined as $L_{\rho'} = \{f \in M \mid \rho'(f) < \infty\}$.

REMARKS 8. The following are some useful facts about dual spaces:

   i) ρ' is a function norm with SFP and $L_{\rho'}$ is a Banach function space.

   ii) There are no unfriendly sets relative to ρ' .

   iii) There exists a sequence $\{\Omega_n\}$ that is jointly ρ and ρ' admissible.

   iv) Higher duals are similarly defined, e.g.

$$\rho''(f) = \sup \{\int_\Omega |fg| d\mu \mid \rho'(g) \le 1\} .$$

   v) If γ is the constant of theorem 6, then $\gamma\rho(f) \le \rho''(f) \le \rho(f)$ for all $f \in L_\rho$; i.e. L_ρ and $L_{\rho''}$ have the same functions and equivalent norms, and they are identical if and only if ρ has SFP.

   vi) The regular and inverse Hölder inequalities hold in the following forms: If $f \in L_\rho$ and $g \in L_{\rho'}$, then $fg \in L_1$ and $\|fg\|_1 \le \rho(f)\rho'(g)$. Conversely, if $fg \in L_1$ for all $g \in L_{\rho'}$ then $f \in L_\rho$. In fact, the inverse Hölder inequality holds with the weakened hypothesis that $fg \in L_1$ for all g in a closed norm-determining subspace of $L_{\rho'}$. This is not stated explicitly in [19] but is readily seen in the proof that appears there, which is also the proof found in [18].

vii) L_ρ, is isometrically isomorphic to a closed sub-space of L_ρ^* (the conjugate space of L_ρ), since if $g \in L_\rho$, then $G(f) = \int_\Omega fg d\mu$ defines $G \in L_\rho^*$ such that $\|G\| = \rho'(g)$. This subspace is not in general all of L_ρ^* . The correspondence is also a lattice isomorphism.

DEFINITION 9. An element $f \in L_\rho$ has <u>absolutely continuous norm</u> if

a) $\lim_n \rho(\chi_{E_n}) = 0$ for any sequence $\{E_n\}$ such that $E_n \subseteq E$ with $\rho(\chi_E) < \infty$ and $\lim_n \mu(E_n) = 0$, and

b) $\lim_n \rho(f\chi_{\Omega - \Omega_n}) = 0$ for any admissible sequence $\{\Omega_n\}$.

Let $L_\rho^\alpha = \{f \in L_\rho \mid f$ has absolutely continuous norm$\}$. Note that L_ρ^α is a closed linear subspace and a normal sublattice of L_ρ

DEFINITION 10. Let π be an admissible sequence $\{\Omega_n\}$. Define $L_\rho^\pi = \overline{sp} \{f \in L_\rho \mid f$ is bounded and has support in some $\Omega_n\}$.

Note that for each π , L_ρ^π is a closed linear subspace and a sublattice of L_ρ .

REMARKS 11. i) It can be shown that $\bigcap_\pi L_\rho^\pi = L_\rho^\alpha$.

ii) Although L_ρ^α is not necessarily norm determining, each L_ρ^π is. (A subspace $Y \subseteq L_\rho$ is <u>norm determining</u> if $\rho'(g) = \sup \{\int |fg| d\mu \mid \rho(f) \le 1, f \in Y\}$ for all $g \in L_\rho$, .)

iii) $(L_\rho^\alpha)^* \cong L_\rho$, if and only if there is an admissible sequence π such that $L_\rho^\pi = L_\rho^\alpha$.

DEFINITION 12. Define $\Sigma_0 = \{E \in \Sigma \mid \rho(\chi_E) < \infty\}$
$$\Sigma_0' = \{E \in \Sigma \mid \rho'(\chi_E) < \infty\} .$$

A $\underline{\text{partition}}$ $\mathcal{E}$ is defined to be a finite disjoint collection of non μ-null members of Σ_o which are of finite measure. Define the $\underline{\text{"averaged" step function}}$ of $f \in L_\rho$ to be

$$f_{\mathcal{E}} = \sum \left(\int_{E_i} |f| \, d\mu / \mu(E_i) \right) \chi_{E_i} \, .$$

DEFINITION 13. A function norm is said to have $\underline{\text{property (J)}}$ if, for each partition $\mathcal{E}$, $\rho(f_{\mathcal{E}}) \leq \rho(f)$.

LEMMA 14. $\underline{\text{If the collection of partitions is regarded as}}$ $\underline{\text{partially ordered by refinement, and if}}$ ρ $\underline{\text{has (J), then}}$ $\rho(f_{\mathcal{E}})$ $\underline{\text{is an increasing function of}}$ $\mathcal{E}$.

Proof. Let $\mathcal{E} = \{E_i | i=1,\ldots,n\}$ and $\mathcal{F} = \{F_j | j=1,\ldots,m\}$ be two partitions such that $\mathcal{E}$ is finer than $\mathcal{F}$; i.e. each E_i is either contained in some F_j or disjoint from all F_j , and each F_j is the union of a collection of the E_i . Renumbering, if necessary, we define $k_o,\ldots,k_m$ such that $k_o = 0$; $E_{k_{j-1}+1},\ldots,E_{k_j} \subseteq F_j$ and

$$\bigcup_{i=k_{j-1}+1}^{k_j} E_i = F_j \quad \text{for} \quad j=1,\ldots,m \; ; \text{ and } E_{k_m+1},\ldots,E_n \text{ are disjoint}$$

from all F_j . A simple computation yields that $(f_{\mathcal{E}})_{\mathcal{F}} = f_{\mathcal{F}}$ and thus $\rho(f_{\mathcal{F}}) = \rho\big((f_{\mathcal{E}})_{\mathcal{F}}\big) \leq \rho(f_{\mathcal{E}})$ by (J).   QED

REMARK 15. i) Property (J) is satisfied by all the well known Banach function spaces, e.g. all the Orlicz spaces (and in particular the Lebesgue spaces).

ii) Property (J) is similar to a property, called $\underline{\text{levelling}}$, in [7]; it was remarked in [7] that if a function norm is levelling then it satisfies (J). No use will be made of this remark below.

There are some properties of function norms with property (J) that are needed and which are proved in [7] under the levelling hypothesis. Since the arguments are virtually the same in the two contexts they will only be sketched here.

LEMMA 16. $\underline{If}$ ρ $\underline{has}$ (J), $\underline{then}$

a) $\rho(\chi_E) \geq \dfrac{\mu(E)}{\mu(E \cup F)} \rho(\chi_F)$ $\underline{for}$ E,F $\underline{of\ finite\ measure}$,

b) $\underline{if}$ $0 < \mu(E) < \infty$ $\underline{then}$ $0 < \rho(\chi_E) < \infty$,

c) $\underline{if}$ $\mu(E) < \infty$ $\underline{then}$ $\mu(E) = \rho(\chi_E) \rho'(\chi_E)$.

Proof. a) If $f = \chi_{E_1}$ and if $\mathcal{E}$ is the one set partition $\{E \cup F\}$, where E and F are of finite measure, then

$$\rho(\chi_E) \geq \rho\left(\frac{\mu(E)}{\mu(E \cup F)} \chi_{E \cup F}\right) = \frac{\mu(E)}{\mu(E \cup F)} \rho(\chi_{E \cup F}) \geq \frac{\mu(E)}{\mu(E \cup F)} \rho(\chi_F) .$$

b) Since ρ is a norm on equivalence classes of functions differing on μ-null sets, $\mu(E) = 0$ if and only if $\rho(\chi_E) = 0$. Assume $\mu(E) < \infty$. Then taking F such that $0 < \mu(F) < \infty$ and $\rho(\chi_F) < \infty$ we have $\rho(\chi_E) \leq \dfrac{\mu(E \cup F)}{\mu(F)} \rho(F) < \infty$ by part a.

c) If $\mu(E) = 0$, the statement is trivial. Let $0 < \mu(E) < \infty$. If g is a simple function in L_ρ , say $g = \sum\limits_{i=1}^{n} \alpha_i \chi_{E_i}$, and if $\mathcal{E}$ is the one set partition $\{E \cap \bigcup\limits_{i=1}^{n} E_i\}$, then $\int_E |g_{\mathcal{E}}| d\mu = \int_E |g| d\mu$.

Moreover $\rho(g_{\mathcal{E}}) \leq \rho(g)$ by (J). Thus,

$$\rho'(\chi_E) = \sup \left\{ \int_E |g_{\mathcal{E}}| d\mu \ \Big|\ \rho(g) \leq 1 \ \text{and}\ g \ \text{simple} \right\}$$

$$= \sup \left\{ \int_E k d\mu \ \Big|\ \rho(k \chi_E) \leq 1 \right\}$$

$$= \mu(E) \rho(\chi_E)^{-1} . \qquad\qquad \text{QED.}$$

COROLLARY 17. **If** ρ **has** (J), **then** $\Sigma_o \cap \Sigma_F = \Sigma_o' \cap \Sigma_F$, **where**
$\Sigma_F = \{E \in \Sigma \mid \mu(E) < \infty\}$. **In particular, every partition has all**
its members in both Σ_o **and** Σ_o' .

COROLLARY 18. **If** ρ **has** (J) **and** $\mu(\Omega) = \infty$, **then** $\mu(E) = \infty$
implies that $\rho(\chi_E) = \rho(\chi_\Omega)$ **(which may be finite or infinite)**.

Proof. There exists an increasing sequence $\{E_n\}$ such that
$\mu(E_n) < \infty$ and $\lim_n \mu(E_n) = \mu(E) = \infty$. Take $\{\Omega_m\}$ to be a ρ-admissible sequence. Then

$$\rho(\chi_E) \ge \rho(\chi_{E_n}) \ge \frac{\mu(E_n)}{\mu(\Omega_m \cup E_n)} \rho(\chi_{\Omega_m}) \qquad \text{for } m,n=1,2,\ldots$$

by lemma 16 (a). Thus

$$\rho(\chi_E) \ge \sup_{m,n} \left[\frac{\mu(E_n)}{\mu(E_n \cup \Omega_m)} \rho(\chi_{\Omega_m}) \right] = \rho(\chi_\Omega) \ .$$

The reverse inequality is always true.   QED.

LEMMA 19. **If** ρ **has property** (J), **then** ρ' **has property** (J).

Proof. Let $f \in L_\rho$ and $h \in L_{\rho'}$. Then

$$\int |fh_{\mathcal{E}}| \, d\mu = \int |f| \sum_{\mathcal{E}} \left(\int_{E_i} |h| \, d\mu / \mu(E_i) \right) \chi_{E_i} \, d\mu$$

$$= \sum_{\mathcal{E}} \left(\int_{E_i} |h| \, d\mu \right) \left(\int_{E_i} |f| \, d\mu \right) / \mu(E_i)$$

$$= \int \sum_{\mathcal{E}} \left(\int_{E_i} |h| \, d\mu / \mu(E_i) \right) \left(\int_{E_i} |f| \, d\mu / \mu(E_i) \right) \chi_{E_i} \, d\mu$$

$$= \int \left[\sum_{\mathcal{E}} \left(\int_{E_i} |h| \, d\mu / \mu(E_i) \right) \chi_{E_i} \right] \left[\sum_{\mathcal{E}} \left(\int_{E_i} |f| \, d\mu / \mu(E_i) \right) \chi_{E_i} \right] d\mu$$

$$= \int |f_{\mathcal{E}} h_{\mathcal{E}}| \, d\mu \ .$$

Similarly, $\int |f_{\mathcal{E}} h| \, d\mu = \int |f_{\mathcal{E}} h_{\mathcal{E}}| \, d\mu$, and therefore $\int |f_{\mathcal{E}} h| \, d\mu = \int |fh_{\mathcal{E}}| \, d\mu$.

Consequently

$$\rho'(h_{\mathcal{E}}) = \sup\left\{\int |f_{\mathcal{E}}h|\,d\mu \,\Big|\, \rho(f) \le 1\right\}$$

$$= \sup\left\{\int |f_{\mathcal{E}}h|\,d\mu \,\Big|\, \rho(f_{\mathcal{E}}) \le 1\right\} \text{ since } \rho(f_{\mathcal{E}}) \le \rho(f) \text{ by (J)}$$

$$\le \rho'(h) .$$

Thus ρ' has property (J).   QED.

CHAPTER II: REPRESENTATION OF LINEAR OPERATORS

1. The Structure of the Space $\mathcal{B}(\mathfrak{X}, L_\rho)$. Let $\mathfrak{X}$ be a Banach space and L_ρ a Banach function space, as defined earlier. It will be assumed that ρ' (and hence ρ'') has property (J). Thus partitions using sets from Σ_0' and those using sets from Σ_0 are identical.

In this section, the continuous linear operators from $\mathfrak{X}$ to L_ρ will be determined in terms of certain (finitely) additive set functions. Note that if $\mathcal{E}$ is a partition and $f = \sum_{\mathcal{E}} \alpha_i \chi_{E_i}$ is a simple function taking its constant values on the members of $\mathcal{E}$ then f is in both L_ρ and $L_{\rho'}$.

DEFINITION 1. Let $x^*(\cdot) : \Sigma_0' \to \mathfrak{X}^*$ be an additive set function. The $\underline{\rho\text{-variation}}$ of $x^*(\cdot)$ is defined as

$$V_\rho(x^*(\cdot)) = \sup_{\|x\| \le 1} \sup_{\mathcal{E}} \rho\left(\sum_{\mathcal{E}} [x^*(E_i)x/\mu(E_i)] \chi_{E_i} \right) .$$

DEFINITION 2. Define $\mathcal{V}_\rho = \{x^*(\cdot) \,|\, x^*(\cdot) : \Sigma_0' \to \mathfrak{X}^*$, $x^*(\cdot)x$ is countably additive and μ-continuous for each $x \in \mathfrak{X}$, and $V_\rho(x^*(\cdot)) < \infty\}$.

Observe that if $x^*(\cdot) \in \mathcal{V}_\rho$, then $x^*(\cdot)$ is finitely additive (and not in general countably additive), and $x^*(\cdot)$ vanishes on μ-null sets.

THEOREM 3. $\mathcal{V}_\rho$ $\underline{\text{is a linear space and}}$ V_ρ $\underline{\text{is a norm on}}$ $\underline{\text{it}}$.

Proof. Everything except definiteness of the norm follows from straight forward computations. If $V_\rho(x^*(\cdot)) = 0$, then using the one set partition $\{E\}$ we have that

$$\rho((x^*(E)x/\mu(E))\chi_E) = 0 \quad \text{for} \quad E \in \Sigma_o' \cap \{E \mid 0 < \mu(E) < \infty\} ,$$

i.e. $(|x^*(E)x|/\mu(E))\rho(\chi_E) = 0$. Consequently, $x^*(E)x = 0$ for each $x \in \mathfrak{X}$ and each E as above. By countable additivity this becomes true for each $x \in \mathfrak{X}$ and all $E \in \Sigma_o'$. Therefore, $x^*(E) = 0$ in $\mathfrak{X}^*$ for each $E \in \Sigma_o'$, i.e. $x^*(\cdot)$ is the zero function in $\mathfrak{V}_\rho$.   QED.

The following two lemmas will establish most of what is needed for the main theorem of this section.

LEMMA 4. <u>Let</u> $T \in \mathfrak{B}(\mathfrak{X}, L_\rho)$ <u>and define</u> $x^*(\cdot)$ <u>on</u> Σ_o' <u>by</u> $x^*(E)x = \int_E (Tx)(\omega)d\mu(\omega)$ <u>for</u> $E \in \Sigma_o'$. <u>Then</u> $x^*(\cdot) \in \mathfrak{V}_\rho$ <u>and</u> $V_\rho(x^*(\cdot))$ $\gamma^{-1}\|T\|$.

Proof. That $x^*(E)x$ is well defined follows by an application of the Hölder inequality; namely,

$$|x^*(E)x| = |\int_E (Tx)(\omega)d\mu(\omega)| \leq \rho(Tx)\rho'(\chi_E) < \infty \quad \text{for} \quad E \in \Sigma_o' .$$

It's clear that $x^*(E)$ is a linear functional on $\mathfrak{X}$. Moreover,

$$\|x^*(E)\|_{\mathfrak{X}^*} = \sup_{\|x\|\leq 1} |x^*(E)x| \leq \sup_{\|x\|\leq 1} \rho(Tx)\rho'(\chi_E) = \|T\|\rho'(\chi_E) < \infty ,$$

so that $x^*(E) \in \mathfrak{X}^*$ for each $E \in \Sigma_o'$.

From the properties of the integral $x^*(\cdot)x$ is countably additive and μ-continuous for each $x \in \mathfrak{X}$. Thus, it remains only to show that $V_\rho(x^*(\cdot)) < \infty$. Since ρ'' has property (J) , then for any partition $\mathcal{E}$ we have

$$\rho''((Tx)_{\mathcal{E}}) = \rho''(\sum_{\mathcal{E}} (x^*(E_i)x/\mu(E_i))\chi_{E_i}) \leq \rho''(Tx) \ .$$

Therefore
$$V_\rho(x^*(\cdot)) = \sup_{\|x\|\leq 1} \ \sup_{\mathcal{E}} \ \rho(\sum_{\mathcal{E}} (x^*(E_i)x/\mu(E_i))\chi_{E_i})$$

$$\leq \sup_{\|x\|\leq 1} \gamma^{-1} \rho(Tx)$$

$$= \gamma^{-1}\|T\| \ . \qquad \text{QED.}$$

Let $x^*(\cdot) \in \mathcal{V}_\rho$. Then for each $x \in \mathfrak{X}$, the scalar valued set function $x^*(\cdot)x$ is μ-continuous and countably additive on Σ_o' , which (in general) is a ring. Since $x^*(\cdot)x$ is finite on Σ_o' and since Σ is the smallest σ-field (in fact, the smallest σ-ring) containing Σ_o' , $x^*(\cdot)x$ has a σ-finite μ-continuous extension to all of Σ (as a signed measure, in general). Thus, the Radon-Nikodym theorem [4] is applicable to the extension of $x^*(\cdot)x$ and guarantees the existence of a unique (a.e.) finite valued measurable function f on Ω such that $\int_E fd\mu$ is the value of the extension for each $E \in \Sigma$. In particular, $x^*(E)x = \int_E fd\mu$ for all $E \in \Sigma_o'$. We denote f by $dx^*(\cdot)x/d\mu$ even though it is actually the Randon-Nikodym derivative of the extension of $x^*(\cdot)x$. With this nomenclature, we have the following

LEMMA 5. **Let** $x^*(\cdot) \in \mathcal{V}_\rho$. **If** T **is defined by** $Tx = dx^*(\cdot)x/d\mu$, **then** $T \in \mathcal{B}(\mathfrak{X}, L_\rho)$.

Proof. The operator T is defined on all $\mathfrak{X}$ and is clearly linear. Moreover, $\int_E Txd\mu = \int_E (dx^*(\cdot)x/d\mu)d\mu = x^*(E)x$ for $E \in \Sigma_o'$. Thus $(Tx)\chi_E \in L_1$ for $E \in \Sigma_o'$ and $\|(Tx)\chi_E\|_1$

$= |x^*(E)x|$. Consequently, if $g = \Sigma \; \alpha_i \chi_{E_i}$ is a ρ' simple function, then $(Tx)g \in L_1$ and

$$\|(Tx)g\|_1 = \left\|(Tx)\left(\sum_i \alpha_i \chi_{E_i}\right)\right\|_1$$

$$\leq \sum_i |\alpha_i| \; (|x^*(E_i)x|/\mu(E_i))\mu(E_i)$$

$$= \int \sum_i |\alpha_i| \; (|x^*(E_i)x|/\mu(E_i)) \chi_{E_i} d\mu$$

$$\leq \rho'(g) \; \rho\left(\sum_i (x^*(E_i)x/\mu(E_i))\chi_{E_i}\right)$$

$$\leq \rho'(g)V_\rho(x^*(\cdot))\|x\| \; .$$

Now, define $M_{\rho'} = \overline{\mathbf{sp}}\{f \in L_{\rho'} | f \text{ simple}\}$ and note that $M_{\rho'}$ is closed and norm determining (II.2.1-3). For any $g \in M_{\rho'}$, there exists a sequence of simple functions $\{g_n\}$ such that $|g_n| \uparrow |g|$ and $\rho'(g-g_n) \to 0$. Then, for a fixed $x \in \mathfrak{X}$, $\|(Tx)g_n\|_1 \leq \rho'(g_n)V_\rho(x^*(\cdot))\|x\| \leq \rho'(g)V_\rho(x^*(\cdot))\|x\|$, i.e. the sequence $\{(Tx)g_n\} \subseteq L_1$ with $\|(Tx)g_n\|_1 \leq M < \infty$ and $|(Tx)g_n| \uparrow |(Tx)g|$. By the Lebesque monotone convergence theorem, $(Tx)g \in L_1$ and $\|(Tx)g_n\|_1 \uparrow \|(Tx)g\|_1$. Thus $(Tx)g \in L_1$ for all g in a closed and norm determining subspace of $L_{\rho'}$. By an application of I.2.7(vi), the "inverse" Hölder inequality, it follows that $Tx \in L_\rho$.

It remains to show that T is bounded. Consider $(Tx)_{\mathcal{E}} = \sum_{\mathcal{E}} (x^*(E_i)x/\mu(E_i))\chi_{E_i}$. If $g \in L_{\rho'}$, then by the proof of lemma I.2.19 we have that

$$\int |g| \, |(Tx)_{\mathcal{E}}| d\mu = \int |g_{\mathcal{E}}| \; |(Tx)_{\mathcal{E}}| d\mu$$

$$\leq \int |g| \ |Tx| \ d\mu < \infty \ .$$

But $\int |g| \ |Tx| d\mu = \int |g| [\lim_{\mathcal{E}} \sum_{\mathcal{E}} (|x^*(E_i)x|/\mu(E_i)) \chi_{E_i}] d\mu$. Consequently, $\lim_{\mathcal{E}} \int |g| \ |(Tx)_{\mathcal{E}}| d\mu \leq \int |g| \lim_{\mathcal{E}} |(Tx)_{\mathcal{E}}| d\mu$. Note that the limit on the left exists since the net is increasing as $\mathcal{E}$ is refined and the terms are bounded above by $\|g \cdot Tx\|_1$ which is finite.

If g is a simple function then equality holds in the last inequality as can be seen by the following computation. Let $g = \sum_j \beta_j \chi_{F_j}$ be in $L_{\rho'}$. Then

$$\int |g| \ \lim_{\mathcal{E}} |(Tx)_{\mathcal{E}}| d\mu = \int [\sum_j |\beta_j| \chi_{F_j}] \ |dx^*(\cdot)x/d\mu| d\mu$$

$$= \int \sum_j |\beta_j| \chi_{F_j} \ |dx^*(\cdot)x|$$

$$= \sum_j |\beta_j| \ |x^*(F_j)x| \ .$$

On the other hand,

$$\lim_{\mathcal{E}} \int |g| \ |(Tx)_{\mathcal{E}}| d\mu = \lim_{\mathcal{E}} \sum_i \sum_j |\beta_j| (\int_{E_i} |Tx| d\mu/\mu(E_i)) \mu(E_i \cap E_j)$$

$$= \lim_{\mathcal{E}} \sum_j \sum_i |\beta_j| (|x^*(E_i)x|/\mu(E_i)) \mu(E_i \cap E_j)$$

Since $\{F_j\}$ is fixed, the partition $\mathcal{E}$ is eventually finer than $\{F_j\}$ and thus $\lim_{\mathcal{E}} \sum_i (|x^*(E_i)x|/\mu(E_i)) \mu(E_i \cap F_j) = |x^*(F_j)x|$. Hence, $\lim_{\mathcal{E}} \int |g| \ |(Tx)_{\mathcal{E}}| d\mu = \sum_j |\beta_j| \ |x^*(F_j)x|$ and there is equality as claimed for a simple function.

By taking the supremum over the simple functions in $L_{\rho'}$

(which are norm determining for $L_{\rho''}$), it follows that

$$\rho''(Tx) = \rho''(\lim_{\mathcal{E}} (Tx)_{\mathcal{E}}) = \sup_{\rho'(g)\leq 1} \int \lim_{\mathcal{E}} |g| \ |(Tx)_{\mathcal{E}}| d\mu$$

$$= \sup_{\rho'(g)\leq 1} \lim_{\mathcal{E}} \int |g| \ |(Tx)_{\mathcal{E}}| d\mu \quad \text{(by the above)}$$

$$= \sup_{\rho'(g)\leq 1} \lim_{\mathcal{E}} \int |g| \sum_{\mathcal{E}} (|x^*(E_i)x|/\mu(E_i)) \chi_{E_i} d\mu \ .$$

Define (temporarily) for the fixed $x^*(\cdot) \in \mho_\rho$ that is under discussion the expression

$$I(g,\mathcal{E}) = \int |g| \sum_{\mathcal{E}} (|x^*(E_i)x|/\mu(E_i)) \chi_{E_i} d\mu \ ,$$

where $g \in L_\rho$, and $\mathcal{E}$ is a partition. Let $x = \sup_{\rho'(g)\leq 1} \sup_{\mathcal{E}} I(g,\mathcal{E})$. Either $\alpha < \infty$ or $\alpha = \infty$.

In the first case, given $\epsilon > 0$ there exists a function g_0 in L_ρ' and a partition $\mathcal{E}_0$ such that $\rho'(g_0) \leq 1$ and $I(g_0,\mathcal{E}_0) \geq \alpha - \epsilon$. Thus $\lim_{\mathcal{E}} I(g_0)_{\mathcal{E}},\mathcal{E}) \geq I((g_0)_{\mathcal{E}_0},\mathcal{E}_0) = I(g_0,\mathcal{E}_0) \geq \alpha - \epsilon$, (where the inequality is true since for a fixed g , $I((g)_{\mathcal{E}},\mathcal{E})$ increases as $\mathcal{E}$ becomes finer). Since ρ' has property (J) $\rho'((g_0)_{\mathcal{E}}) \leq \rho''(g_0) \leq 1$. Hence $\sup_{\rho'(g)\leq 1} \lim_{\mathcal{E}} I(g,\mathcal{E}) \geq \alpha - \epsilon$.

Since $\epsilon > 0$ was arbitrary, we have $\sup_{\rho'(g)\leq 1} \lim_{\mathcal{E}} I(g,\mathcal{E}) \geq \alpha$.

In the other case, when $\alpha = \infty$, then for each N there exist $g_0 \in L_\rho$, and a partition $\mathcal{E}_0$ such that $\rho'(g_0) \leq 1$ and $I(g_0,\mathcal{E}_0) \geq N$. The same reasoning as in the first case leads to the conclusion that $\sup_{\rho'(g)\leq 1} \lim_{\mathcal{E}} I(g,\mathcal{E}) \geq N$ for every

$N > 0$ and hence that $\sup_{\rho'(g)\leq 1} \lim_{\mathcal{E}} I(g,\mathcal{E}) = \infty$. But this is a contradiction since $\displaystyle\sup_{\rho'(g)\leq 1}\ \lim_{\mathcal{E}}\ I(g,\mathcal{E}) = \rho''(Tx) < \infty$.

Consequently, $\alpha < \infty$ and

$$\sup_{\rho'(g)\leq 1}\ \sup_{\mathcal{E}}\ I(g,\mathcal{E}) \leq \sup_{\rho'(g)\leq 1}\ \lim_{\mathcal{E}}\ I(g,\mathcal{E})$$

Since the reverse inequality is always true, equality holds.

Thus $\rho''(Tx) = \displaystyle\sup_{\mathcal{E}}\ \rho''(\textstyle\sum_{\mathcal{E}}\ (|x^*(E_i)x|/\mu(E_i))\chi_{E_i})$, which implies

that

$$\gamma\rho(Tx) \leq \sup_{\mathcal{E}}\ \rho(\textstyle\sum_{\mathcal{E}}\ (|x^*(E_i)x|/\mu(E_i))\chi_{E_i})$$

Taking the supremum over the unit ball of $\mathfrak{X}$ leads to the

desired result that $\|T\| \leq \gamma^{-1}V_\rho(x^*(\cdot))$.   QED.

THEOREM 6. <u>If</u> ρ' <u>satisfies condition</u> (J), <u>then there</u>
<u>is an isomorphism between</u> $\mathcal{B}(\mathfrak{X},L_\rho)$ <u>and</u> $\mathcal{V}_\rho$ <u>given by</u>

$$x^*(E)x = \int Tx\ d\mu$$

<u>and</u> $Tx = dx^*(\cdot)x/d\mu$.

<u>Moreover</u> $\gamma\|T\| \leq V_\rho(x^*(\cdot)) \leq \gamma^{-1}\|T\|$ <u>for corresponding elements</u>
<u>where</u> γ <u>is the constant appearing in I.2.6</u>.

Proof. The two preceding lemmas have established this
theorem once one notes the simple fact that the mappings
$x^*(\cdot) \to T$ and $T \to x^*(\cdot)$ given in the theorem are linear,
1-1, onto, and inverse to each other.   QED.

COROLLARY 7. <u>If</u> ρ' <u>has property</u> (J) <u>and the strong</u>
<u>Fatou property, then</u> $\mathcal{B}(\mathfrak{X},L_\rho) \cong \mathcal{V}_\rho$.

This is immediate since ρ has SFP if and only if $\gamma = 1$.

COROLLARY 8. $\mathcal{V}_\rho$ <u>is complete under the norm</u> V_ρ .

REMARKS 9. Results for a similar problem may also be
found in [7], in which arbitrary measure spaces and vector
valued functions were allowed. The type of function norm
considered there, however, is more restrictive. In fact, the
main results in [7] are proved under the hypotheses (in addition to levelling) that ρ has SFP, that (certain) step
functions are dense in L_ρ , and that ρ has certain "con-
tinuity" properties.

2. The Structure of the Space $\mathscr{B}(M_\rho,\mathfrak{X})$. This section
should deal with the characterization of $\mathscr{B}(L_\rho,\mathfrak{X})$, but the
problem appears to be intractable in this generality. The
results that will be presented are for a closed subspace of L_ρ
In Chapter III the special case of characterization of the
space L_ρ^* will be presented.

DEFINITION 1. Let $M_\rho = \overline{sp}\ \{f \in L_\rho | f$ is bounded and has
support in $\Sigma_0\}$.

It is clear that M_ρ is a closed subspace which is normal
and a sublattice. It will be shown that it is norm determining.
(In many cases, it is true that $M_\rho = L_\rho$; e.g. an Orlicz space
L^Φ with Φ satisfying the Δ_2 condition [17,28,36].)

LEMMA 2. $M_\rho = \overline{sp}\ \{f \in L_\rho | f$ simple$\}$.

(Note that, since μ is σ-finite, if $0 \le f \in M_\rho$ then
there is a sequence of simple functions $\{f_n\}$ such that
$0 \le f_n \uparrow f$ and $\rho(f-f_n) \to 0$. Recall that a ρ-simple function
is of the form $\Sigma \alpha_i \chi_{E_i}$ with $\{E_i\}$ being a finite, disjoint

collection of finite measure members of Σ_0 .)

THEOREM 3. $\underset{\pi}{\cup} L_\rho^\pi \subseteq M_\rho$. <u>Moreover, if either</u> i) $\mu(\Omega) < \infty$, <u>or ii)</u> $\mu(\Omega) = \infty$, ρ' <u>has</u> (J), <u>and</u> $\rho(\chi_\Omega) = \infty$, <u>then</u> $\underset{\pi}{\cup} L_\rho^\pi = M_\rho$.

Proof. By definition, $L_\rho^\pi = \overline{sp}\ \{f \in L_\rho | f$ is bounded and has support inside a member of $\pi\}$ where π is an admissible sequence. Since every member of an admissible sequence is a member of Σ_0 it follows that $L_\rho^\pi \subseteq M_\rho$ for each π and hence that $\underset{\pi}{\cup} L_\rho^\pi \subset M_\rho$.

For the case of equality, consider first i) $\mu(\Omega) < \infty$. By lemma 2, there exists for each $f \in M_\rho$ a sequence $\{f_k\}$ such that

$$f_k = \sum_{i=1}^{n_k} \alpha_i^k \chi_{E_i^k} \quad \text{and} \quad \rho(f-f_k) \to 0 ,$$

where for each k , $\{E_i^k\}_{i=1}^{n_k}$ are disjoint and in Σ_0 . Construct $\Omega_1 = \overset{n_1}{\underset{i=1}{\cup}} E_i^1$. Then $\Omega_1 \in \Sigma_0$. Having constructed $\Omega_1, \ldots, \Omega_{m-1}$, we construct $\Omega_m = \Omega_{m-1} \cup \overset{n_m}{\underset{i=1}{\cup}} E_i^m$ and $\Omega_m \in \Sigma_0$ for $m = 1, 2, \ldots$. Note that $\{\Omega_m\}$ is an increasing sequence. Now take $\pi : V_m \uparrow \Omega$ to be any admissible sequence and define $\Omega_m^o = \Omega_m \cup V_m$. Then $\Omega_m^o \uparrow \Omega$ and $\rho(\chi_{\Omega_m^o}) + \rho(\chi_{V_m}) < \infty$. If we define π_o to be the sequence $\{\Omega_m^o\}$, then $f \in L_\rho^{\pi_o}$ and equality holds in this case.

Next we consider case ii) where ρ' has (J) and $\mu(\Omega) = \rho(\chi_\Omega) = \infty$. Then ρ'' also has (J). By Corollary I.2.18 if $\mu(E) = \infty$, then $\rho''(\chi_E) = \rho''(\chi_\Omega) = \infty$ because $\rho(\chi_\Omega) = \infty$.

Thus, if $\rho(\chi_E) < \infty$ then $\mu(E) < \infty$; i.e. members of Σ_0 have finite measure in this case. Consequently, the above construction of a specific admissible sequence goes through.   QED.

That equality may not hold in the event that (Ω, Σ, μ) is σ-finite, ρ has (J), and $\rho(\chi_\Omega) < \infty$ is readily seen by observing the case of L_∞ on the real line with Lebesgue measure. In this case, $\rho(\chi_\Omega) = 1$. The function $u(\omega) \equiv 1$ is certainly a member of M_ρ but is not in any L_ρ^π and hence u cannot be approximated in the L_∞ norm by elements of L_ρ^π .

As in the previous section, $\mathfrak{X}$ is a Banach space and it is assumed that ρ' has (J). Consequently Σ_0 partitions and Σ_0' partitions are identical.

DEFINITION 4. The $\underline{\rho'\text{-variation}}$ of an $\mathfrak{X}$-valued set function is $W_{\rho'}(x(\cdot)) = \sup_{\|x^*\|\leq 1} \sup_{\mathcal{E}} \rho'\left(\sum_{\mathcal{E}} (x^*x(E_i)/\mu(E_i))\chi_{E_i}\right)$.

Define $\mathbb{w}_{\rho'} = \{x(\cdot) \,|\, x(\cdot):\Sigma_0 \to \mathfrak{X} \,,\, x(\cdot)$ is finitely additive, $x(\cdot)$ vanishes on μ-null sets, and $W_{\rho'}(x(\cdot)) < \infty\}$.

THEOREM 5. $\mathbb{w}_{\rho'}$ <u>is a linear space and</u> $W_{\rho'}$ <u>is a norm on it</u>.

In the characterization of $\mathfrak{B}(M_\rho, \mathfrak{X})$ it will be desirable to integrate members of M_ρ against set functions in $\mathbb{w}_{\rho'}$. In order to do this, Bartle's treatment of integration [2] will be used. In the context of the present work the definition is as follows:

DEFINITION 6. A function $f \in M$ is integrable over Ω

with respect to an $\mathfrak{X}$-valued finitely additive set function $x(\cdot)$ if there is a sequence $\{f_n\}$ of simple functions such that

$\quad$ i) $f_n \to f$ in $x(\cdot)$ measure

$\quad$ ii) $\lambda_n(\cdot)$ are uniformly absolutely continuous

and $\quad$ iii) $\lambda_n(\cdot)$ are equicontinuous,

where $\lambda_n(E) = \int_E f_n dx$ for all $E \in \Sigma$.

(The integral of a simple function is, as usual, taken to be

$$\int_E (\sum_{i=1}^{n} \alpha_i \chi_{E_i}) dx = \sum_{i=1}^{n} \alpha_i x(E_i \cap E)) .$$

REMARK 7. i) If f is $x(\cdot)$ integrable, then for each $E \in \Sigma$ the limit of the indefinite integrals $\int_E f_n dx$ exists in norm and is denoted as $\lambda(E) = \int_E f dx$; i.e. $\lambda(E) = \lim_n \lambda_n(E)$ uniformly in the strong topology of $\mathfrak{X}$.

ii) In definition 6, the concepts of uniform absolute continuity and of equicontinuity are with respect to the semi-variation, $\|x\|(E) = \sup \{\| \sum_{\mathcal{E}} \alpha_i x(E_i) \| \mid |\alpha_i| \leq 1, E_i \subseteq E\}$, rather than the variation, $|x|(E) = \sup \{\sum_{\mathcal{E}} \|x(E_i)\| \mid E_i \subseteq E\}$, because $x(\cdot)$ is vector valued. Note that [6] $\|x\|(E) \leq |x|(E)$, but the reverse need not be true. If, however, $\mathfrak{X}$ is the scalar field, then they coincide.

LEMMA 8. <u>Let</u> $g = \sum_{i=1}^{k} \beta_i \chi_{G_i}$ <u>be a simple function in</u> M_ρ <u>and let</u> $x(\cdot) \in \mathfrak{w}_{\rho'}$. <u>Denote the indefinite integral by</u>

$$\sigma(E) = \int_E g dx = \sum_{i=1}^{k} \beta_i x(E \cap G_i) \quad , \underline{for} \ E \in \Sigma .$$

Then $\|\sigma(E)\|_{\mathfrak{x}} \leq \rho(g\chi_E)W_{\rho},(x(\cdot))$

    Proof. For any x^* in the unit ball of $\mathfrak{x}^*$,

$$|x^*\sigma(E)| = |x^* \int_E gdx|$$

$$= |x^* \sum_{i=1}^{k} \beta_i x(G_i \cap E)|$$

$$= |x^* \sum_{i=1}^{k} \beta_i (x(G_i \cap E)/\mu(G_i \cap E))\mu(G_i \cap E)|$$

$$= |\int_E x^* \sum_{i=1}^{k} \beta_i (x(G_i \cap E)/\mu(G_i \cap E))\chi_{G_i \cap E}d\mu|$$

$$= |\int_E [\sum_{i=1}^{k} \beta_i \chi_{G_i \cap E}][\sum_{i=1}^{k} (x^*x(G_i \cap E)/\mu(G_i \cap E))\chi_{G_i \cap E}]d\mu|$$

$$\leq \int_E |g\chi_E| \sum_{i=1}^{k} (|x^*x(G_i \cap E)|/\mu(G_i \cap E))\chi_{G_i \cap E}d\mu$$

$$\leq \rho(g\chi_E)\rho'(\sum_{i=1}^{k}(|x^*x(G_i \cap E)|/\mu(G_i \cap E))\chi_{G_i \cap E})$$

$$\leq \rho(g\chi_E)W_{\rho},(x(\cdot)) .$$

Thus $|x^*\sigma(E)| \leq \rho(g\chi_E)W_{\rho},(x(\cdot))$ for each x^* in the unit ball
of $\mathfrak{x}^*$. Therefore $\|\sigma(E)\|_{\mathfrak{x}} \leq \rho(g\chi_E)W_{\rho},(x(\cdot))$.   QED

    LEMMA 9. **Let** g, $x(\cdot)$, **and** σ **be as in lemma 8. Then**
$\|\sigma(E)\|_{\mathfrak{x}} \leq \|x\|(E) \|g\|_{\infty}$.

    Proof. Excluding the trivially true case that $\|g\|_{\infty} = 0$,
one has

$$\|\sigma(E)\|_{\mathfrak{x}} = \|\sum_{i=1}^{k} \beta_i x(G_i \cap E)\|_{\mathfrak{x}}$$

$$= \|\sum_{i=1}^{k} (\beta_i/\|g\|_{\infty})x(G_i \cap E)\|_{\mathfrak{x}} \|g\|_{\infty}$$

$$\leq \sup_{\mathcal{E}} \{\|\sum \alpha_i x(E_i)\|_{\mathfrak{x}} \|g\|_{\infty} \big| E_i \subseteq E, |\alpha_i| \leq 1\}$$

$$= \|x\|(E) \, \|g\|_\infty \, . \qquad\qquad \text{QED}$$

LEMMA 10. **If** $x(\cdot) \in \mathbb{w}_\rho$, **then** $\|x\|(E) \le W_{\rho'}(x(\cdot))\rho(\chi_E)$.

Proof. Let $\mathcal{E} = \{E_i\}_{i=1}^n$ be a partition of sets in E and let $|\alpha_i| \le 1$. Then

$$\left\| \sum_\mathcal{E} \alpha_i x(E_i) \right\|_{\mathfrak{X}} = \sup_{\|x^*\| \le 1} \left| x^* \sum_\mathcal{E} \alpha_i x(E_i) \right|$$

$$= \sup_{\|x^*\| \le 1} \left| \int \sum_\mathcal{E} (\alpha_i x^* x(E_i)/\mu(E_i)) \chi_{E_i} d\mu \right|$$

$$= \sup_{\|x^*\| \le 1} \left| \int \left[\sum_\mathcal{E} (x^* x(E_i)/\mu(E_i)) \chi_{E_i} \right] \left[\sum_\mathcal{E} \alpha_i \chi_{E_i} \right] d\mu \right|$$

$$\le \sup_{\|x^*\| \le 1} \rho'\left(\sum_\mathcal{E} (x^* x(E_i)/\mu(E_i)) \chi_{E_i} \right) \rho(\Sigma \alpha_i \chi_{E_i}) \, .$$

Moreover, since $\{E_i\}$ are disjoint and $|\alpha_i| \le 1$, $\rho(\sum_\mathcal{E} \alpha_i \chi_{E_i}) \le \rho(\chi_E)$
Thus $\left\| \sum_\mathcal{E} \alpha_i x(E_i) \right\|_{\mathfrak{X}} \le \sup_{\|x^*\| \le 1} \rho'\left(\sum_\mathcal{E} (x^* x(E_i)/\mu(E_i)) \chi_{E_i} \right) \rho(\chi_E)$. Con-

sequently, $\|x\|(E) = \sup \left\{ \left\| \sum_\mathcal{E} \alpha_i x(E_i) \right\|_{\mathfrak{X}} \middle| E_i \subseteq E, \, |\alpha_i| \le 1 \right\}$

$$\le W_{\rho'}(x(\cdot))\rho(\chi_E) \, . \qquad\qquad \text{QED.}$$

These lemmas will be used to establish a key result of this section.

THEOREM 11. **If** $f \in M_\rho$ **and** $x(\cdot) \in \mathbb{w}_{\rho'}$, **then** f **is** $x(\cdot)$ **integrable**.

Proof. We prove the theorem for $f \ge 0$ and observe that the general case follows in the usual fashion by linearity or the decomposition of f into positive functions. Since $0 \le f \in M_\rho$, there exists a sequence of simple functions $\{f_n\}$ such that $0 \le f_n \uparrow f$ and $\rho(f-f_n) \to 0$. We denote their integrals as

$\lambda_n(E) = \int f_n d\mu$. What needs to be established are the three criteria of definition 6.

       i) Let $T_n^\alpha = \{\omega \mid f(\omega) - f_n(\omega) \geq \alpha\}$ for $\alpha > 0$. Since the $\{f_n\}$ are increasing, $n \geq m$ implies that $T_n^\alpha \subseteq T_m^\alpha$. Consequently $\|x\|(T_n^\alpha)\downarrow$ as $n \to \infty$. Moreover by lemma 10, $\|x\|(T_n^\alpha) \leq W_\rho ,(x(\cdot))\rho(\chi_{T_n^\alpha}) \leq W_\rho ,(x(\cdot))\rho(\chi_{T_1^\alpha}) < \infty$.

Hence, for each $\alpha > 0$, $\|x\|(T_n^\alpha)$ converges as $n \to \infty$. Denote the limit by S_α . If $S_\alpha > 0$, then for any ϵ , such that $0 < \epsilon < 1$, there exists N_1 with $0 \leq \|x\|(T_n^\alpha) - S_\alpha < \epsilon$, for $n \geq N_1$. There also exists N_2 such that $\rho''(f-f_n) > \alpha S_\alpha / W_\rho ,(x(\cdot))$ for $n \geq N_2$. Take $N_0 = \max \{N_1, N_2\}$. Now, by definition of $\|x\|(\cdot)$, there exist $\{E_i^\circ\}$, a partition of $T_{n_0}^\alpha$, and scalars $\{\alpha_i^\circ\}$ with $|\alpha_i^\circ| \leq 1$ such that

$$S_\alpha + \epsilon > \|\sum_i \alpha_i^\circ x(E_i^\circ)\|_{\mathfrak{x}} > S_\alpha .$$

But then $\rho''(f - f_{n_0}) = \sup \{\int |h| \; |f - f_{n_0}| d\mu \mid \rho'(h) \leq 1\}$

$$\geq \int |f - f_{n_0}| \cdot |(\sum_i (\alpha_i^\circ x^* x(E_i^\circ)/\mu(E_i^\circ))\chi_{E_i}) / W_\rho ,(x(\cdot))| d\mu ,$$

for $\|x^*\| \leq 1$,

$$\geq (\alpha / W_\rho ,(x(\cdot))) \int |\sum_i (\alpha_i^\circ x^* x(E_i))/\mu(E_i^\circ)\chi_{E_i^\circ}| d\mu ,$$

since $E_i^\circ \subseteq T_{n_0}^\alpha$,

$$= (\alpha / W_\rho ,(x(\cdot))) \sum_i |\alpha_i x^* x(E_i^\circ)| , \text{ since } \{E_i^\circ\} \text{ disjoint},$$

$$\geq (\alpha / W_\rho ,(x(\cdot))) |x^* \sum_i \alpha_i^\circ x(E_i^\circ)| , \text{ for any } \|x^*\| \leq 1 .$$

The Hahn-Banach Theorem ensures the existence of $x_0^* \in \mathfrak{x}^*$

such that $\|x_o^*\| = 1$ and $|x_o^* \sum_i \alpha_i^o x(E_i^o)| = \|\sum_i \alpha_i^o x(E_i^o)\|_{\bar{x}}$.

Consequently $\rho''(f-f_{n_o}) \geq \alpha \|\sum_i \alpha_i^o x(E_i^o)\|_{\bar{x}}/W_\rho'(x(\cdot))$

$$\geq \alpha S_\alpha/W_\rho'(x(\cdot)) .$$

But, on the other hand, $\rho''(f-f_{n_o}) \leq \rho(f-f_{n_o}) < \alpha S_\alpha/W_\rho'(x(\cdot))$.
This is a contradiction. Hence $S_\alpha = 0$ for each $\alpha > 0$;
i.e. $\|x\|(T_n^\alpha) \to 0$ as $n \to \infty$ for any $\alpha > 0$. Thus $f_n \to f$
in $x(\cdot)$ measure.

      ii) Let $\epsilon > 0$ be fixed. What is to be shown
is that there exists $\delta > 0$ such that $\|x\|(E) < \delta$ implies
$\|\lambda_n(E)\|_{\bar{x}} < \epsilon$ for $n = 1,2,\ldots$. Lemma 9 guarantees that
$\|\lambda_n(E)\| \leq \|x\|(E) \|f_n\|_\alpha$ for each n . Picking $\delta_n = \epsilon/2\|f_n\|_\infty$
gives $\|\lambda_n(E)\| < \epsilon/2$ whenever $\|x\|(E) < \delta_n$. Lemma 8 states
that

$$\|\lambda_n(E)\|_{\bar{x}} \leq \rho(f_n \chi_E) W_\rho'(x(\cdot))$$

and $\quad\quad\quad\quad \|\lambda_n(E) - \lambda_m(E)\|_{\bar{x}} \leq \rho((f_n - f_m)\chi_E) W_\rho'(x(\cdot))$

$$\leq \rho(f_n - f_m) W_\rho'(x(\cdot)) , \text{ for } E \in \Sigma .$$

Moreover, there exists N such that $\rho(f_n - f_m) < \epsilon/2W_\rho'(x(\cdot))$
for $n,m \geq N$. Finally, we note that $\|\lambda_n(E)\| \leq |\lambda_n(E) - \lambda_N(E)\|$
$+ \|\lambda_N(E)\|$. Now if $n \geq N$ and $\|x\|(E) \leq \delta_N$, then $\|\lambda_n(E)\|$
$< \epsilon/2 + \epsilon/2 = \epsilon$. Thus, for $n \geq N$, $\|x(E)\| \leq \delta_N$ implies
$\|\lambda_n(E)\| < \epsilon$. Consequently, if $\delta = \min(\delta_1,\ldots,\delta_N)$ and
$\|x\|(E) < \delta$ then $\|\lambda_n(E)\| < \epsilon$ for all n .

      iii) Let $\epsilon > 0$ be fixed. To be established is
the existence of $E_\epsilon \in \Sigma$ with finite semivariation such
that, if $G \in \Sigma$ and $G \subseteq \Omega - E_\epsilon$, then $\|\lambda_n(G)\| < \epsilon$ for $n=1,2,\ldots$.

There does exist an index N such that $\rho(f-f_n) < \epsilon/W_{\rho'}(x(\cdot))$
for all $n \geq N$. Let E_ϵ be the union of the supports of the
functions f_n for $n = 1,2,\ldots,N$. Then, by lemma 10
$\|x\|(E_\epsilon) \leq W_{\rho'}(x(\cdot))\rho(\chi_{E_\epsilon}) < \infty$, whereas by lemma 8
$\|\lambda_n(G)\| \leq \rho(f_n\chi_G)W_{\rho'}(x(\cdot))$ for all n and each $G \in \Sigma$.
Consequently, if $G \subseteq \Omega-E_\epsilon$, then

$$\begin{aligned}
\|\lambda_n(G)\| &\leq \rho(f_n\chi_G)W_{\rho'}(x(\cdot)) \\
&\leq \rho(f\chi_G)W_{\rho'}(x(\cdot)) \\
&\leq [\rho((f-f_N)\chi_G) + \rho(f_N\chi_G)]W_{\rho'}(x(\cdot)) .
\end{aligned}$$

But $\rho(f_N\chi_G) = 0$. Hence

$$\begin{aligned}
\|\lambda_n(G)\| &\leq \rho((f-f_N)\chi_G)W_{\rho'}(x(\cdot)) \\
&\leq \rho(f-f_N)W_{\rho'}(x(\cdot)) \\
&< \epsilon \qquad \text{for } n = 1,2,\ldots \; . \qquad \text{QED.}
\end{aligned}$$

We will denote the integral of $f \in M_\rho$ with respect to
$x(\cdot) \in \mathbb{w}_{\rho'}$ by $\int f dx$ (Note that if $x(\cdot)$ is scalar valued
and countably additive, this integral coincides with the Lebesgue
integral.).

The next theorem is the main one of this section.

THEOREM 12. _If_ ρ' _has_ (J), _then_ $\mathbb{B}(M_\rho,\mathfrak{X})$ _and_ $\mathbb{w}_{\rho'}$
are linearly homeomorphic under the correspondence $Tf = \int f dx$
for $f \in M_\rho$ _and_ $x(E) = T(\chi_E)$ _for_ $E \in \Sigma_0$, _where_
$\|T\| \leq W_{\rho'}(x(\cdot)) \leq \gamma^{-1}\|T\|$ _for corresponding elements_.

Proof. Given $x(\cdot) \in \mathbb{w}_{\rho'}$, define $Tf = \int f dx$ for
$f \in M_\rho$. This is well defined by theorem 11, and $T:M_\rho \to \mathfrak{X}$.

It is clear that T is linear. Moreover, by lemma 8, if h is simple then $\|\int h\,dx\| \leq \rho(h)W_{\rho'}(x(\cdot))$. Since the simple functions are dense in M_ρ ,

$$\|T\| = \sup \{\|Th\| \mid \rho(h) \leq 1 \text{ , } h \text{ simple}\}$$
$$= \sup \{\|\int h\,dx\| \mid \rho(h) \leq 1 \text{ , } h \text{ simple}\}$$
$$\leq W_{\rho'}(x(\cdot)) \ .$$

Thus $T \in \mathcal{B}(M_\rho, \mathfrak{X})$ and $\|T\| \leq W_{\rho'}(x(\cdot))$.

Conversely, if $T \in \mathcal{B}(M_\rho, \mathfrak{X})$, define $x(\cdot):\Sigma_0 \to \mathfrak{X}$ by $x(E) = T(\chi_E)$. Clearly $x(\cdot)$ is additive and vanishes on μ-null sets. All that remains is the computation of $W_{\rho'}(x(\cdot))$. Let $\mathcal{E}$ be a partition, $x^* \in \mathfrak{X}^*$ with $\|x^*\| \leq 1$, $h = \sum_j \beta_j \chi_{F_j}$ be a simple function, and c_i be scalars such that $|\alpha_i| = 1$ and $\alpha_i x^* x(E_i) = |x^* x(E_i)|$. Then

$$\int |h| \sum_{\mathcal{E}} (|x^* x(E_i)|/\mu(E_i)) \chi_{E_i} \, d\mu = \sum_i \alpha_i [\sum_j (|\beta_j|\mu(E_i \cap F_j)/\mu(E_i))x^* x(E_i)]$$

$$= x^*(\sum_i \alpha_i [\sum_j (|\beta_j|\mu(E_i \cap F_j)/\mu(E_i)x(E_i)])$$

$$\leq \|\int g\,dx\|_{\mathfrak{X}} \ ,$$

where $g = \sum_i \alpha_i (\int_{E_i} |h|\,d\mu/\mu(E_i)\chi_{E_i}$ is a simple function with

$\rho''(g) \leq \rho''(h)$ by property (J). Consequently $\gamma\rho(g) \leq \rho(h)$, and

$$\rho'(\sum_{\mathcal{E}}(|x^* x(E_i)|/\mu(E_i))\chi_{E_i}) = \sup \{\int |h| \sum_{\mathcal{E}}(|x^* x(E_i)|/\mu(E_i))\chi_{E_i}\,d\mu \mid \rho(h) \leq 1 \text{ ,}$$

$$h \text{ simple}\}$$

$$\leq \sup \{\|\int g\,dx\|_{\mathfrak{X}} \mid \gamma\rho(g) \leq 1 \text{ , } g \text{ simple}\}$$

$$= \gamma^{-1} \|T\| \; .$$

Hence, $x(\cdot) \in \mathbb{w}_\rho{}'$ and $W_\rho{}'(x(\cdot)) \leq \gamma^{-1}\|T\|$.

The maps $\mathcal{B}(M_\rho, \mathfrak{X}) \to \mathbb{w}_\rho{}'$ and $\mathbb{w}_\rho{}' \to \mathcal{B}(M_\rho, \mathfrak{X})$ are clearly isomorphisms and are inverses to each other. QED.

COROLLARY 13. <u>If</u> ρ <u>has SFP</u>, $\mathcal{B}(M_\rho, \mathfrak{X}) \cong \mathbb{w}_\rho{}'$.

A natural question to ask, concerning members of $\mathbb{w}_\rho{}'$, is when are $x(\cdot)$ μ-continuous? A partial answer is in the following.

LEMMA 14. <u>If</u> $M_\rho = L_\rho^\alpha$, <u>then each</u> $x(\cdot) \in \mathbb{w}_\rho{}'$ <u>is</u> μ-<u>continuous</u>.

Proof. Taking the one set partition E , we have that $\rho'((x^* x(E)/\mu(E))\chi_E) \leq \|x^*\| W_\rho{}'(x(\cdot))$. Hence, $\|x(E)\| \leq (\mu(E)/\rho'(\chi_E))W_\rho{}'(x(\cdot)) = \rho''(\chi_E)W_\rho{}'(x(\cdot))$ by lemma I.2.16. But $M_\rho = L_\rho^\alpha$ if and only if $M_{\rho''} = L_{\rho''}^\alpha$; therefore, if $\chi_E \in M_{\rho''}$ then $\chi_E \in L_{\rho''}^\alpha$, and hence $\lim\limits_{\mu(E) \to 0} \rho''(\chi_E) = 0$. Thus,

$$\lim\limits_{\mu(E) \to 0} \|x(E)\| = 0 \; .\qquad\qquad \text{QED.}$$

As one might expect, the condition $M_\rho = L_\rho^\alpha$ is equivalent to the condition $\lim\limits_{\mu(E) \to 0} \rho(\chi_E) = 0$ which has been noted to be equivalent to $\lim\limits_{\mu(E) \to 0} \mu(E)/\rho''(\chi_E) = 0$. Although each of these suffices to imply that all members of $\mathbb{w}_\rho{}'$ are μ-continuous, the necessity is somewhat more complicated. In the case that $\mathfrak{X}$ is the scalars, then the condition is necessary as will be seen in the next chapter. In the vector case one needs some compactness conditions. This matter will be discussed elsewhere.

CHAPTER III: CHARACTERIZATIONS OF L_ρ^* .

In this chapter two characterization of L_ρ^* , the space
of continuous linear functionals on a Banach function space
L_ρ , will be presented. One of these will be more general
than the other (since property (J) will not be assumed) but
less useful (due to the nature of the norm computations of
the set functions involved). Both characterizations will
proceed by use of the factor space $(L_\rho/M_\rho)^*$. In this chapter
the usual sign for isometric isomorphism, $\cong$, will denote,
in addition to linear isomorphism, lattice isomorphism.

1. The Structure of the Quotient Space N_ρ and Its
Conjugate. The point of view adopted in this section will
be to work with real vector spaces since the complex cases are
easily obtainable by noting that each complex normed vector
lattice is the complexification of a real space with the same
properties.

DEFINITION 1. Define $N_\rho = L_\rho/M_\rho$ and denote the canon-
ical map as $\lambda:L_\rho \to N_\rho$ where $\lambda f = f + M_\rho$

Since M_ρ is a closed subspace of L_ρ , general facts
on quotient spaces (e.g. [4] or [6]) give us that N_ρ is a
Banach space with norm

$$\|\hat{f}\| = \inf \{\rho(f) | f \in \hat{f}\} \text{ where } \hat{f} = \lambda f .$$

Moreover, λ is continuous, interior, homomorphic, and has norm ≤ 1. In addition, $\lambda^*: N_\rho^* \to M_\rho^\perp$ is an isometric isomorphic surjection, where $M_\rho^\perp = \{x^* \in L_\rho^* \mid x^*(f) = 0$ for all $f \in M_\rho\}$.

Since M_ρ is a sublattice and lattice ideal in L_ρ , N_ρ has the induced order $\hat{f} \geq \hat{\theta}$ if there is $f_0 \in \hat{f}$ such that $f_0 \geq g_0$ for some $g_0 \in M_\rho$. Moreover, it is easily seen that N_ρ is an AB-lattice and that λ is a lattice homomorphism. Finally, N_ρ^* is a boundedly complete AB-lattice since it is the conjugate space of an AB-lattice (cf. [4]).

DEFINITION 2. Let $\widetilde{L}_\rho = \{f \mid f = \bigvee_{i=1}^{n} f_i$, $f_i \geq 0$, $\rho(f_i) \leq 1$, $1 \leq n < \infty\}$.

Note that $\widetilde{L}_\rho$ is convex and norm determining in L_ρ and that $\widetilde{L}_\rho$ is a sublattice. Moreover, since λ is interior, $\lambda(\widetilde{L}_\rho)$ contains the nonnegative elements of the open unit ball of N_ρ . In order to use the tools developed in [29] and [1], it is here that we need to assume on ρ the following

Property (I). $\lambda(\widetilde{L}_\rho)$ lies in the closed unit ball of N_ρ .

REMARK 3. Let $\{\Omega_m\}$ be an arbitrary ρ-admissible sequence. Then for $0 \leq f \in L_\rho$, define $\{f_n\}$ by

$$f_n(\omega) = \begin{cases} f(\omega) & \text{if } \omega \in \{\omega \mid |f(\omega)| \leq n\} \cap \Omega_n \\ 0 & \text{otherwise.} \end{cases}$$

Then $f_n \in M_\rho$ and $f_n \uparrow f$; equivalently $\bar{f}_n = f - f_n \downarrow 0$ and $\tilde{f}_n \in \lambda(f)$. It is clear that the assumption of property (I) for ρ is equivalent to assuming that $\lim_n \rho(\tilde{f}_n) \leq 1$ for all $f \in L_\rho$.

The well known Banach function spaces have property (I); e.g. the Orlicz spaces [17, 29, 1] and the modulared normed function spaces [25].

LEMMA 4. *If* ρ *has property* (I), *then for* $0 \leq z^* \in N_\rho^*$ *we have* $\|z^*\| = \sup \{|z^*(\hat{f})| \mid \hat{f} \in \lambda(\tilde{L}_\rho)\}$.

LEMMA 5. *If* ρ *has property* (I), *then* N_ρ^* *is an abstract* L *space in the sense of Kakutani* [9], *hereinafter abbreviated as an AL-space; i.e.,* N_ρ^* *is an AB-lattice with the additional property that if* z_1^* *and* z_2^* *are non-negative members of* N_ρ^* *then* $\|z_1^* + z_2^*\| = \|z_1^*\| + \|z_2^*\|$.

Proof. The proof is similar to that for Orlicz spaces [29]. Since $\|z_1^* + z_2^*\| \leq \|z_1^*\| + \|z_2^*\|$ is true in any event, it remains to prove the opposite inequality. Let z_1^* and z_2^* be non-negative members of N_ρ^* and let $\epsilon > 0$ be given. Then, there exist $\hat{f}_1$ and $\hat{f}_2$, non-negative, in $\lambda(\tilde{L}_\rho)$ such that $\|z_i^*\| - \epsilon/2 \leq z_i^* \hat{f}_i$, $i=1,2$. Consequently,

$$\|z_1^*\| + \|z_2^*\| - \epsilon \leq z_1^* \hat{f}_1 + z_2^* \hat{f}_2 \leq (z_1^* + z_2^*)(\hat{f}_1 \vee \hat{f}_2) = (z_1^* + z_2^*)(\widehat{f_1 \vee f_2}) .$$

Moreover, $\widehat{f_1 \vee f_2} \in \lambda(\tilde{L}_\rho)$. Thus by property (I) and lemma 4, $\|\widehat{f_1 \vee f_2}\| \leq 1$ and $\|z_1^*\| + \|z_2^*\| - \epsilon \leq \|z_1^* + z_2^*\|$. Since $\epsilon > 0$ was arbitrary, the result follows.   QED

Some facts about N_ρ^* follow from the general theory of

AL-spaces [4, 9, 10]. Denote by j the canonical map from N_ρ into N_ρ^{**}. Since N_ρ^{*} is an AL-space, N_ρ^{**} is a boundedly complete abstract M space in the sense of Kakutani (abbreviated as AM-space); i.e., an AB-lattice with the additional property that $0 \leq f$, $g \in N_\rho^{**}$ implies that $\|f \vee g\| = \max(\|f\|, \|g\|)$. Moreover, the positive cone of N_ρ^{**} has an interior point u_o such that the unit ball of N_ρ^{**} is precisely $\{x^{**} \in N_\rho^{**} \mid -u_o \leq x^{**} \leq u_o\}$. Hence $N_\rho^{**} \subseteq C(S)$, the space of continuous functions on the compact Hausdorff space S formed by the weak* topologized set of extreme points on the positive face of the unit sphere in N_ρ^{***} (where the correspondent in $C(S)$ for u_o is the function identically 1). But any closed linear sublattice of $C(S)$ is always an AM-space; hence identifying under the linear isometry-lattice homomorphism of N_ρ^{**} with $C(S)$, we find that N_ρ is an AM-space. This general description will serve as motivation for the characterization of N_ρ^{*}, which will, in a sense, identify N_ρ inside $L_\infty(\Omega, \Sigma, \mu)$.

DEFINITION 6. Denote by $ba(\Omega, \Sigma, \mu)$ the collection of bounded additive set functions on Σ which vanish on μ-null sets. The space $ba(\Omega, \Sigma, \mu)$ is to be endowed with the usual [6, 35] norm and order; with these it is an AB-lattice. The norm is the variation

$$\|v\| = \sup \left\{ \sum_{i=1}^{n} |v(E_i)| \ \middle| \ \{E_i\} \leq \Sigma \text{ and disjoint} \right\}.$$

The order is "set wise," i.e. $v \geq 0$ if $v(E) \geq 0$ for all $E \in \Sigma$. Note that the lattice operations are given by

$$(v_1 \wedge v_2)(E) = \inf\{(v_1(T)+v_2(E \cap T^c)) \mid T \in \Sigma , \; T \subseteq E\}$$

and $(v_1 \vee v_2)(E) = \sup\{(v_1(T)+v_2(E \cap T^c)) \mid T \in \Sigma , \; T \subseteq E\}$. Standard theorems about $ba(\Omega,\Sigma,\mu)$, e.g. Jordan decomposition, are found in [6] and [35].

DEFINITION 7. A non-negative set function $v \in ba(\Omega,\Sigma,\mu)$ is called _purely finitely additive_ [35] if for every countably additive ψ with $0 \leq \psi \leq v$ it results that $\psi = 0$. A general v is purely finitely additive if its decomposition yields positive purely finitely additive set functions.

We name $ca(\Omega,\Sigma,\mu) = \{v \in ba(\Omega,\Sigma,\mu) \mid v$ countably additive$\}$ and $pfa(\Omega,\Sigma,\mu) = \{v \in ba(\Omega,\Sigma,\mu) \mid v$ purely finitely additive$\}$. It is well known (and easily demonstrated [35, 6]) that $ca(\Omega,\Sigma,\mu)$ and $pfa(\Omega,\Sigma,\mu)$ are closed linear subspaces and sublattices of $ba(\Omega,\Sigma,\mu)$ and that $ba(\Omega,\Sigma,\mu) = ca(\Omega,\Sigma,\mu) \oplus pfa(\Omega,\Sigma,\mu)$.

LEMMA 8. Let $v \in ba(\Omega,\Sigma,\mu)$. If there exists a sequence $\{B_n\}$ in Σ such that $\lim_n \psi(B_n) = 0$ for any non-negative $\psi \in ca(\Omega,\Sigma,\mu)$ and $v(B_n) = v(\Omega)$, n=1,2..., then $v \in pfa(\Omega,\Sigma,\mu)$.

This result with proof is given in [35] and is quoted here for use below.

THEOREM 9. Let $z^* \in N_\rho^*$. If $z^* \geq 0$, define $v(E) = \|z_E^*\|$; for general z^* , define $v(E) = \|(z_E^*)^+\| - \|(z_E^*)^-\|$. Then $v \in pfa(\Omega,\Sigma,\mu)$. Moreover, the correspondence $z^* \to v$ is a linear isometry and a lattice isomorphism.

Notation. Given $z^* \in N_\rho^*$, define $z_E^*(\hat{f}) = z^*(\hat{f}\chi_E)$ for $\hat{f} \in N_\rho$. It is clear that $z_E^* \in N_\rho^*$.

Proof. If $z^* \geq 0$, then we define $v(E) = \|z_E^*\|$ and v is non-negative, vanishes on μ-null sets and is bounded by $\|z^*\|$. Additivity follows from Theorem 5. For general z^* , we define $v(E) = \|(z_E^*)^+\| - \|(z_E^*)^-\|$ and again have that v vanishes on μ-null sets, is bounded by $\|z^*\|$, and is additive. Thus $v \in ba(\Omega,\Sigma,\mu)$. The fact that the map is linear and an isometry is immediate. Moreover, since positive elements (and only positive elements) in N_ρ^* have positive correspondents in $ba(\Omega,\Sigma,\mu)$, the map is a lattice homomorphism (and hence an isomorphism due to the isometry).

All that remains is to show that $v \in pfa(\Omega,\Sigma,\mu)$. It suffices to consider $z^* \geq 0$. Given $\epsilon > 0$, there exists $\hat{f}_\epsilon \geq 0$ such that $z^*(\hat{f}_\epsilon) \leq \|z^*\| \leq z^*(\hat{f}_\epsilon)+\epsilon$ and $\|\hat{f}_\epsilon\| = 1$. Introduce the auxiliary set function $v_\epsilon(E) = z^*(f_\epsilon\chi_E)$. Clearly $v_\epsilon \in ba(\Omega,\Sigma,\mu)$ and

$$v_\epsilon(E) = z^*(f_\epsilon\chi_E) = z_E^*(\hat{f}_\epsilon) \leq \|z_E^*\| = v(E) .$$

Moreover, by the choice of $\hat{f}_\epsilon$, $v_\epsilon(\Omega) \leq v(\Omega) \leq v_\epsilon(\Omega)+\epsilon$. Hence, for $E \in \Sigma$,

$$v(E)+v(E^C) \leq v_\epsilon(E)+v_\epsilon(E^C)+\epsilon \; ;$$

i.e., $$0 \leq v(E) \leq v_\epsilon(E)+[v_\epsilon(E^C) - v(E^C)]+\epsilon .$$

The term in brackets is non-positive. Consequently

$$v_\epsilon(E) \leq v(E) \leq v_\epsilon(E)+\epsilon , \quad \text{for all } E \in \Sigma .$$

Therefore, for each $\epsilon > 0$, there exists v_ϵ such that

$\|v-v_\epsilon\| < \epsilon$. Since $pfa(\Omega,\Sigma,\mu)$ is closed, it will follow that $v \in pfa(\Omega,\Sigma,\mu)$ once it is shown that $v_\epsilon \in pfa(\Omega,\Sigma,\mu)$.

Case 1. Let $\mu(\Omega) < \infty$. Take $f_0 \in \hat{f}_\epsilon$ such that $f_0 \geq 0$. Define $F_k = \{\omega \mid k-1 \leq f_0(\omega) < k\}$ for $k=1,2,\dots$. Note that $\overset{\infty}{\underset{k=1}{\cup}} F_k = \Omega$. Define $B_n = \overset{\infty}{\underset{k=n}{\cup}} F_k$ and note that the sequence $\{B_n\}$ is decreasing. Since $f_\epsilon \chi_{\underset{k=1}{\overset{n-1}{\cup}} F_k} \in \hat{\theta} = \lambda(M_\rho)$, we have

$$v_\epsilon(B_n) = z^*(f_\epsilon \chi_{\underset{k=n}{\overset{\infty}{\cup}} F_k}) + z^*(f_\epsilon \chi_{\underset{k=1}{\overset{n-1}{\cup}} F_k} = z^*(\hat{f}_\epsilon) = v_\epsilon(\Omega) .$$

On the other hand, $\overset{\infty}{\underset{k=1}{\cup}} F_k = \Omega$ and $\mu(\Omega) < \infty$ imply that $\mu(B_n) = \mu(\overset{\infty}{\underset{k=n}{\cup}} F_k) \downarrow 0$ as $n \to \infty$. $\psi \in ca(\Omega,\Sigma,\mu)$ must be μ-continuous, hence $\psi(B_n) \downarrow 0$ for all $\psi \in ca(\Omega,\Sigma,\mu)$. By lemma 3, $v_\epsilon \in pfa(\Omega,\Sigma,\mu)$ and hence $v \in pfa(\Omega,\Sigma,\mu)$.

Case 2. Let (Ω,Σ,μ) be σ-finite. Then there is a ρ-admissible sequence $\Omega_m \uparrow \Omega$ with $\rho(\chi_{\Omega_m}) < \infty$ and $\mu(\Omega_m) < \infty$. Letting $\Sigma_m = \Sigma \mid \Omega_m$, we note that $z^*_{\Omega_m} \in N^*_\rho(\Omega_m,\Sigma_m,\mu)$ and that $v_m \in ba(\Omega_m,\Sigma_m,\mu)$ can be defined as arising from this linear functional. In fact, $v_m(E) = v(E \cap \Omega_m) = v(E)$ for $E \in \Sigma_m$. Case 1 guarantees that $v_m \in pfa(\Omega_m,\Sigma_m,\mu)$. Now write $v = v_c + v_p$ where $0 \leq v_c \leq v$, $0 \leq v_p \leq v$, $v_c \in ca(\Omega,\Sigma,\mu)$ and $v_p \in pfa(\Omega,\Sigma,\mu)$. But $v_m = v_c \mid_{\Omega_m} + v_p \mid_{\Omega_m}$ and $v_c \mid_{\Omega_m} \leq v \mid_{\Omega_m}$, so that $v_c \mid_{\Omega_m} = 0$ and $v \mid_{\Omega_m} = v_p \mid_{\Omega_m}$. It follows that

$$v_c(E) = (v_c(E) - v_c(E \cap \Omega_m)) + v_c(E \cap \Omega_m) \to 0 \text{ as } n \to \infty ,$$

Since $\underset{m}{\lim}\, v_c(E \cap \Omega_m) = v_c(E)$ and $v_c(E \cap \Omega_m) = v_c \mid_{\Omega_m}(E) = 0$.

In other words $\nu_c(E) = 0$, $E \in \Sigma$ and thus $\nu \in \text{pfa}(\Omega,\Sigma,\mu)$.   QED

In order to effect the reverse map from $\text{pfa}(\Omega,\Sigma,\mu)$ to N_ρ^* a type of integration of elements of N_ρ with respect to members of $\text{pfa}(\Omega,\Sigma,\mu)$ is needed. The difficulty is that although $\nu \in \text{pfa}(\Omega,\Sigma,\mu)$ is defined on the σ-field Σ in Ω , the members of N_ρ are not point functions on Ω . The integration used here is a variant of that found in [30] and [32].

DEFINITION 10. Let $0 \le \nu \in \text{pfa}(\Omega,\Sigma,\mu)$ and $0 \le f \in L_\rho$. Then we define $I_\nu(f) = \inf\{ \sum_{i=1}^{n} \|\hat{f}\chi_{E_i}\| \nu(E_i) \,|\, \{E_i\} \text{ disjoint finite partition of } \Omega\}$.

THEOREM 11. <u>If</u> $0 \le \nu$, $\psi \in \text{pfa}(\Omega,\Sigma,\mu)$, $0 \le f \in L_\rho$, $a,b \ge 0$ <u>then</u>

(i) $I_\nu(af) = aI_\nu(f)$.

(ii) $0 \le f \le g$ <u>implies that</u> $I_\nu(f) \le I_\nu(g)$.

(iii) $0 \le I_\nu(f) \le \|\hat{f}\| \nu(\Omega)$.

(iv) $I_\nu(f+g) = I_\nu(f) + I_\nu(g)$.

(v) $I_{a\nu+b\psi}(f) = aI_\nu(f) + bI_\psi(f)$.

Proof. The first three properties are immediate from the definition and (v) is a routine computation. Property (iv) will be shown in a manner similar to that found in [1] and [29].

Case 1. Let $0 \le f, g \in L_\rho$ with f and g having disjoint supports, say $F = \{\omega | f(\omega) > 0\}$ and $G = \{\omega | g(\omega) > 0\}$. Let $\{E_k\}_{k=1}^{n}$ be any finite disjoint collection of sets in Σ , and define

$$C_k = \begin{cases} F \cap E_k & k=1,\ldots,n \\ G \cap E_{k-n} & k=n+1,\ldots,2n \end{cases} .$$

Then $\{C_k\}_{k=1}^{2n} \subseteq \Sigma$ and $f\chi_{C_{k+n}} = g\chi_{C_k} = 0$ for $k=1,2,\ldots,n$.

Thus $(f+g)\chi_{E_k} = f\chi_{C_k} + g\chi_{C_{k+n}}$. Now, recalling that N_ρ is

an **AM**-space and that v is additive, we have

$$\sum_{k=1}^{n} \| \widehat{(f+g)\chi_{E_k}} \| v(E_k) = \sum_{k=1}^{n} \| \widehat{f\chi_{C_k}} + \widehat{g\chi_{C_{k+n}}} \| v(E_k)$$

$$= \sum_{k=1}^{n} \max\{ \| \widehat{f\chi_{C_k}} \| , \| \widehat{g\chi_{C_{k+n}}} \| \}(v(C_k) + v(C_{k-n})) .$$

$$\geq \sum_{k=1}^{n} \| \widehat{f\chi_{C_k}} \| v(C_k) + \sum_{k=n+1}^{2n} \| \widehat{g\chi_{C_k}} \| v(C_k)$$

$$= \sum_{k=1}^{2n} \| \widehat{f\chi_{C_k}} \| v(C_k) + \sum_{k=1}^{2n} \| \widehat{g\chi_{C_k}} \| v(C_k)$$

$$\geq I_v(f) + I_v(g) \quad \text{by definition of } I_v .$$

Taking the infinum on the left gives $I_v(f+g) \geq I_v(f) + I_v(g)$.

Conversely, let $\{D_k\}_{k=1}^{m}$ be another finite disjoint

collection. Let

$$B_k = \begin{cases} F \cap E_k & k=1,\ldots,n \\ G \cap D_k & k=n+1,\ldots,n+m \end{cases} .$$

Then $\displaystyle\sum_{k=1}^{n} \| \widehat{f\chi_{E_k}} \| v(E_k) + \sum_{j=1}^{m} \| \widehat{g\chi_{D_j}} \| v(D_j)$

$$\geq \sum_{k=1}^{n} \| \widehat{f\chi_{E_k \cap F}} \| v(E_k \cap F) + \sum_{j=1}^{m} \| \widehat{g\chi_{D_j \cap G}} \| v(D_j \cap G)$$

$$= \sum_{k=1}^{n+m} \| \widehat{(f+g)\chi_{B_k}} \| v(B_k)$$

$$\geq I_\nu(f+g) \ .$$

Hence $I_\nu(f)+I_\nu(g) \geq I_\nu(f+g)$. Thus, in this case of disjoint supports of f and g , $I_\nu(f)+I_\nu(g) = I_\nu(f+g)$. Moreover, by induction,

$$I_\nu\left(\sum_{k=1}^{n} \alpha_k \widehat{f\chi_{E_k}}\right) = \sum_{k=1}^{n} \alpha_k I_\nu\left(\widehat{f\chi_{E_k}}\right)$$

where $\alpha_k \geq 0$ and $\{E_k\}_{k=1}^{n}$ are disjoint.

Case 2. Let $0 \leq g \leq f$ with $f(\omega) > 0$ a.e. so that $g/f \in L_\infty$. Then, given $\epsilon > 0$, there exists a finite disjoint collection of set $\{E_k\}_{k=1}^{n}$ and non-negative scalars $\{\alpha_k\}_{k=1}^{n}$ such that

$$\sum_{k=1}^{n} \alpha_k \chi_{E_k} \leq \frac{g}{f} \leq \sum_{k=1}^{n} (\alpha_k+\epsilon)\chi_{E_k} \ ;$$

i.e.,
$$\sum_{k=1}^{n} \alpha_k f\chi_{E_k} \leq g \leq \sum_{k=1}^{n} (\alpha_k+\epsilon) f\chi_{E_k} \ .$$

Thus, by case 1, we have

$$I_\nu(f+g) \leq I_\nu\left(f + \sum_{k=1}^{n} (\alpha_k+\epsilon) f\chi_{E_k}\right)$$

$$= \sum_{k=1}^{n} (1+\alpha_k+\epsilon) I_\nu\left(f\chi_{E_k}\right)$$

$$= (1+\epsilon)I_\nu(f) + I_\nu\left(\sum_{k=1}^{n} \alpha_k f\chi_{E_k}\right)$$

$$\leq (1+\epsilon)I_\nu(f) + I_\nu(g) \ .$$

Since $\epsilon > 0$ is arbitrary, $I_\nu(f+g) \leq I_\nu(f)+I_\nu(g)$. Conversely, the opposite inequality is gotten by interchanging the roles of the initial two inequalities.

Case 3. Let $0 \leq f$, $g \in L_\rho$ and $F = \{\omega | f(\omega) \geq g(\omega)\}$ and $G = \{\omega | f(\omega) < g(\omega)\}$. Note that $F \cup G = \Omega$ and $F \cap G = \varphi$. Then,

$$
\begin{aligned}
I_\nu(f+g) &= I_\nu((f\chi_F + g\chi_F) + (f\chi_G + g\chi_G)) \\
&= I_\nu(f\chi_F - g\chi_F) + I_\nu(f\chi_G + g\chi_G) \qquad \text{by case 1} \\
&= I_\nu(f\chi_F) + I_\nu(g\chi_F) + I_\nu(f\chi_G) + I_\nu(g\chi_G) \quad \text{by case 2} \\
&= I_\nu(f\chi_F + f\chi_G) + I_\nu(g\chi_F + g\chi_G) \qquad \text{by case 1} \\
&= I_\nu(f) + I_\nu(g) \ . \qquad\qquad\qquad \text{QED}
\end{aligned}
$$

Extension of definition 10. The integral $I_\nu(f)$ is well defined for $0 \leq \nu \in pfa(\Omega,\Sigma,\mu)$ and $0 \leq f \in L_\rho$. The definition is extended to all of $pfa(\Omega,\Sigma,\mu)$ and L_ρ in the usual manner, i.e. by linearity on the decomposition of general elements into their positive parts. Moreover $I_\nu(f)$ is indistinguishable for all f on the same coset in N_ρ and hence may be regarded as an integral of elements in N_ρ by writing $I_\nu(\hat{f}) = I_\nu(f)$, $f \in \hat{f}$.

COROLLARY 12. *If* $\nu \in pfa(\Omega,\Sigma,\mu)$ *and* $z^*(f) = I_\nu(\hat{f})$, *then* $z^* \in N_\rho^*$.

THEOREM 13. $N_\rho^* \cong P_{\rho'}$, *where* $P_{\rho'}$ *is a closed subspace of* $pfa(\Omega,\Sigma,\mu)$. *The correspondence is given by the following:* *given* $\nu \in P_{\rho'}$, *define* $z^*(\hat{f}) = I_\nu(\hat{f})$ *for* $\hat{f} \in N_\rho$; *conversely,* *given* $z^* \in N_\rho^*$, *define* $\nu(E) = \|(z_E^*)^+\| - \|(z_E^*)^-\|$ *for* $E \in \Sigma$.

Proof. By Theorem 9 there is an isometric linear and lattice isomorphism $C : N_\rho^* \to pfa(\Omega,\Sigma,\mu)$. By Theorem 11, there

is a bounded linear and lattice homomorphism $D : pfa(\Omega,\Sigma,\mu) \to N_\rho^*$.
(Note that, in general, C is not onto and D is not 1-1.)

We first show that $DCz^* = z^*$ for $z^* \in N_\rho^*$ by the following computation (done only for non-negative z^*). Given
$0 \le z^* \in N_\rho^*$, define $v(E) = \|z_E^*\|$ and $y^*(\hat{f}) = I_v(\hat{f})$. Then
for $\{E_i\}$ a disjoint, finite partition of Ω we have

$$\sum_{i=1}^{n} \|\hat{f}\chi_{E_i}\| \, v(E_i) = \sum_{i=1}^{n} \|\hat{f}\chi_{E_i}\| \, \|z_{E_i}^*\|$$

$$\ge \sum_{i=1}^{n} z_{E_i}^* (\hat{f}\chi_{E_i})$$

$$= z^*(\hat{f}) \ .$$

Thus $z^*(\hat{f}) \le \inf \sum_{i=1}^{n} \|f\chi_{E_i}\| \, v(E_i) = y^*(\hat{f})$. Conversely,

$\|y^*\| = |v|(\Omega) \ge \|z^*\|$; and, since $y^*-z^* \ge 0$,

$\|y^*-z^*\| = \|y^*\| - \|z^*\| = 0$. Consequently $y^* = z^*$, and thus
$DCz^* = z^*$.

Define $P = CD$. It is clear that P is a linear,
bounded, lattice homomorphic operator on $pfa(\Omega,\Sigma,\mu)$. In
fact P is a bounded projection since $P^2 = (CD)^2 = C(DC)D = CD = P$.
(It is automatic that $\|P\| \ge 1$. In this case, a slight computation is available to show that $\|P\| = 1$.) Denote the
range of P as $\mathcal{P}_{\rho'}$. Since P is a bounded projection,
$\mathcal{P}_{\rho'}$ is closed. Moreover, by the construction, $\mathcal{P}_{\rho'} \cong N_\rho^*$
under the mapping C to which D is an inverse on $\mathcal{P}_{\rho'}$.   QED .

THEOREM 14. <u>The elements of</u> $\mathcal{P}_{\rho'}$ <u>are those members of</u>
$pfa(\Omega,\Sigma,\mu)$ <u>whose support lies inside the support of a function</u>

<u>in</u> $\widetilde{L}_\rho$ <u>which is not in</u> M_ρ .

Proof. Let C and D be as in the proof of Theorem 13. It is known that $P_{\rho\prime} = \{v \in \text{pfa}(\Omega,\Sigma,\mu) \mid CDv = v\}$ and that in general $\psi = CDv$ then $\psi(E) \leqslant v(E)$ for $0 \leqslant v \in \text{pfa}(\Omega,\Sigma,\mu)$. Since both ψ and v are additive, $\psi = v$ if and only if $\psi(\Omega) = v(\Omega)$. By lemma 8, there exists a decreasing sequence $\{B_n\}$ in Σ such that $\mu(B_n)\downarrow 0$ and $v(B_n) = v(\Omega)$; i.e., the support of v is contained inside each B_n . Construct for each $n=1,2,\ldots$ the disjoint sequence $\{G_j^n\}_{j=1}^{n+1}$ as

$$G_1^1 = B_1^c \quad , \quad G_2^1 = B_1 \quad ;$$

$$G_j^n = G_j^{n-1} \quad , \quad j=1,\ldots,n-1 \quad ;$$

$$G_n^n = B_{n-1}\backslash B_n \quad , \quad G_{n+1}^n = B_n \ .$$

Now, for $\|\hat{f}\| \leqslant 1$, $\psi(\Omega) \geq \inf_{\mathcal{E}} \sum_i \|\hat{f}\chi_{E_i}\|\, v(E_i)$

$$= \lim_{\mathcal{E}} \sum_i \|\hat{f}\chi_{E_i}\|\, v(E_i) \quad ,$$

since refining the partition decreases the value of the sum. Replace the Moore-Smith limit over the net of refinements by the cofinal sequence $\{G_n^j\}$ in order to have

$$\psi(\Omega) \geq \lim_n \sum_i \|\hat{f}\chi_{G_i}\|\, v(G_i^n)$$

$$= \lim_n \|\hat{f}\chi_{B_n}\|\, v(B_n)$$

$$= \lim_n \|\hat{f}\chi_{B_n}\|\, v(\Omega) \ .$$

This last is true for $\|\hat{f}\| \leq 1$. Consequently, $\psi(\Omega) = v(\Omega)$ if and only if there is $\hat{f}_0 \in N_\rho$ such that $\|\widehat{f_0 \chi_{B_n}}\| = \|\hat{f}_0\| = 1$, which occurs if and only if there is $g \in L_\rho$ with $\{\omega | g(\omega) > n\} = B_n$.

Thus, $v \in P_\rho$, if and only if such a g exists. It remains only to show that $g \in \tilde{L}_\rho \setminus M_\rho$. It is clear that $g \notin M_\rho$. Moreover $\|\hat{g}\| = \lim_n \rho(g \chi_{B_n})$. Define $f_0 \in L_\rho$ as $f_0 = g/\|\hat{g}\|$. Then $\{\omega | f_0(\omega) > n/\|\hat{g}\|\} = B_n$ and $\lim_n \rho(f_0 \chi_{B_n}) = \|\hat{f}_0\| = 1$. Therefore, $\|\widehat{f_0 \chi_{B_m}}\| = \lim_n \rho(f_0 \chi_{B_m} \chi_{C_n^m})$, where

$$C_n^m = \{\omega | f_0 \chi_{B_m}(\omega) > n/\|\hat{g}\|\} . \text{ But } f_0 \chi_{B_m} \chi_{C_n^m} = f_0 \chi_{B_n} \text{ for } n \geq m ;$$

hence $\rho(f_0 \chi_{B_m} \chi_{C_n^m}) = \rho(f_0 \chi_{B_n})$ for $n \geq m$. Consequently,

$$\|\widehat{f_0 \chi_{B_m}}\| = \lim_n \rho(f_0 \chi_{B_m} \chi_{C_n^m}) = \lim_n \rho(f_0 \chi_{B_n}) = \|\hat{f}_0\| , \quad m=1,2,\dots .$$

It follows that $g \in \tilde{L}_\rho$.   QED

REMARKS. The observation of Theorem 14 was made in the case of Orlicz spaces in [29]. The use of a projection to determine P_ρ, follows the device used in the case of Orlicz spaces of finitely additive set functions in [33].

2. General Forms of Linear Functionals on L_ρ . In this section two characterizations of the conjugate space of L_ρ will be presented.

Recall that $M_\rho^\perp = \{x^* \in L_\rho^* | x^* f = 0 \text{ for all } f \in M_\rho\}$ is a closed, linear, (lattice) semi-normal subspace of L_ρ^*

since M_ρ is such in L_ρ . We will denote by $(M_\rho^\perp)^{o \cdot c \cdot}$ the
lattice orthogonal complement of $M_\rho^\perp$ in L_ρ^* ; viz,

$$(M_\rho^\perp)^{o \cdot c \cdot} = \{y^* \in L_\rho^* \mid |y^*| \wedge |x^*| = 0 \text{ for all } x^* \in M_\rho^\perp\} \ .$$

THEOREM 1. $L_\rho^* \cong M_\rho^\perp \oplus (M_\rho^\perp)^{o \cdot c \cdot}$. <u>Moreover</u>, $(M_\rho^\perp)^{o \cdot c \cdot} \cong M_\rho^*$.

Proof. This result follows from the general theory of
normed lattices. As in [4] or [26], since L_ρ is an AB-lattice,
L_ρ^* is a boundedly complete (in [26] "universally continuous")
AB-lattice. A theorem in [26] states that in a boundedly
complete AB-lattice any norm-closed, linear, (lattice) semi-
normal subspace is normal. Thus $L_\rho^* \cong M_\rho^\perp \oplus (M_\rho^\perp)^{o \cdot c \cdot}$.

That $(M_\rho^\perp)^{o \cdot c \cdot} \cong M_\rho^*$ follows from a general technique
(e.g. [31]). Because of the direct sum decomposition, it
follows that $L_\rho^*/M_\rho^\perp \cong (M_\rho^\perp)^{o \cdot c \cdot}$. But recall that $(L_\rho/M_\rho)^* \cong M_\rho^\perp$.
If we let i be the injection $i : M_\rho \to L_\rho$, then the adjoint
$i^* : L_\rho^* \to M_\rho^*$ is a continuous (linear and lattice) homomorphism
onto M_ρ^* with kernel $M_\rho^\perp$. Let $\tilde{\lambda}$ be the quotient map
$L_\rho^* \to L_\rho^*/M_\rho^\perp$, and define the map i' so as to make the following diagram commute:

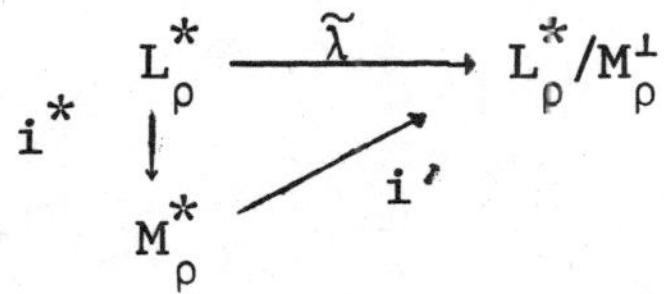

Thus i' is automatically an isometric (linear and lattice)
isomorphism of M_ρ^* and $L_\rho^*/M_\rho^\perp$. Consequently, $M_\rho^* \cong (M_\rho^\perp)^{o \cdot c \cdot}$.

QED

COROLLARY 2. <u>For</u> $x^* \in L_\rho^*$, <u>let</u> $x^* = y^* + z^*$ <u>be the</u> <u>unique decomposition given by Theorem 1, where</u> $y^* \in (M_\rho^\perp)^{o.c.}$ <u>and</u> $z^* \in M_\rho^\perp$. <u>Then</u> $\|x^*\| = \|y^*\| + \|z^*\|$.

Proof. It suffices to consider $0 \le x^* \in L_\rho^*$. Then by Theorem 1 we have $0 \le y^* \in (M_\rho^\perp)^{o.c.}$ and $0 \le z^* \in M_\rho^\perp$ such that $x^* = y^* + z^*$ and $x^* \wedge y^* = 0$. It is always true that $\|x^*\| \le \|y^*\| + \|z^*\|$. Conversely, take $0 \le f \in L_\rho$ with $\rho(f) \le 1$. Then $\|x^*\| \ge x^* f = (y^* + z^*)(f) = (y^* \vee z^*)(f)$, since $y^* \wedge z^* = 0$. But $(y^* \vee z^*)(f) = \sup\{y^*(g) + z^*(h) \mid f = g+h , g \ge 0 , h \ge 0\}$. Thus $\|x^*\| \ge y^*(g) + z^*(h)$ for any $g,h \ge 0$ with $\rho(g+h) \le 1$. Since $y^* \in (M_\rho^\perp)^{o.c.}$, y^* is the norm preserving extension of some element in M_ρ^* . In fact, given $\epsilon > 0$, there exists $g_\epsilon \in M_\rho$ with $\rho(g_\epsilon) \le 1$ such that $y^*(g_\epsilon) \ge \|y^*\| - \epsilon/2$. Similarly, since $z^* \in M_\rho^\perp$, it follows that there is $\hat{z}^* \in N_\rho^*$ such that $z^*(f) = \hat{z}^*(\hat{f})$, where λ is the quotient map $L_\rho \to N_\rho$ and $\hat{f} = \lambda f$. By definition, there exists $\hat{h}_\epsilon \ge \hat{0}$ in N_ρ with $\|\hat{h}_\epsilon\| < 1$ such that $\hat{z}^*(\hat{h}_\epsilon) \le \|\hat{z}^*\| - \epsilon/2$. Consequently, $z^*(h) \ge \|z^*\| - \epsilon/2$ for all $h \in \hat{h}_\epsilon$. Moreover, we note that there exists $h_0 \in \hat{h}_\epsilon$ such that $h_0 \ge 0$ and $\rho(h_0 + g_\epsilon) \le 1$, since $g_\epsilon \in M_\rho$ implies that

$$\|\hat{h}_\epsilon\| = \inf\{\rho(h) \mid h \in \hat{h}_\epsilon\}$$

$$= \inf\{\rho((h - g_\epsilon) + g_\epsilon) \mid h \in \hat{h}_\epsilon\}$$

$$= \inf\{\rho(u + g_\epsilon) \mid u \in \hat{h}_\epsilon\} .$$

For the above choice of g_ϵ and h_0 , we have

$$\|x^*\| \geq y^*(g_\epsilon) + z^*(h_0) \geq \|y^*\| + \|z^*\| - \epsilon .$$

Since $\epsilon > 0$ is arbitrary, the result follows.   QED.

In light of this theorem, all that is needed for the representation of L_ρ^* is a representation of M_ρ^* . If ρ' satisfies (J), then section II.2 gives M_ρ^* as (linear and lattice) isomorphic and topologically equivalent to $w_\rho^R{}'$, the space of real-valued, μ-null, finitely additive set functions on Σ_0 of finite ρ'-variation.

DEFINITION 3. Define $G_{\rho'} = w_{\rho'} \times P_{\rho'}$, with norm $\|(G,v)\| = W_{\rho'}(G) + |v|(\Omega)$ and order $(G,v) \geq 0$ if $G \geq 0$ and $v \geq 0$. (It is clear that $G_{\rho'}$ is an AB-lattice.)

THEOREM 4. <u>If L_ρ has property (I) and ρ' has property (J), then there is a homeomorphic (linear and lattice) isomorphism between L_ρ^* and $G_{\rho'}$ .. Moreover, if ρ has SFP,</u> $L_\rho^* \cong G_{\rho'}$.

The proof of Theorem 4 follows from II.2, III.1.13, III.2.1, and III.2.2.

This theorem can be slightly improved in the sense that the characterization of L_ρ^* can (and will) be given without the assumption that ρ' has (J). This comes from a general characterization of M_ρ^* . (This method leads to a result that in the presence of (J) reduces to the above result in the sense that the spaces of set functions are algebraically the same with equivalent norms.)

We first establish some notation. Let $\pi = \{\Omega_n\}_{n=1}^\infty$ be

a ρ-admissible sequence. Denote by Σ_n the restriction of Σ
to Ω_n . Note that Σ_n coincides with Σ_0 restricted to Ω_n
and that Σ_n is a σ-field in Ω_n . Each function in $M_\rho(\Omega,\Sigma,\mu)$
has a restriction in $M_\rho(\Omega_n,\Sigma_n,\mu)$ and these restrictions may
be regarded as belonging to $M_\rho(\Omega,\Sigma,\mu)$ in the sense that such
an element of $M_\rho(\Omega,\Sigma,\mu)$ vanishes off Ω_n . It will suffice
to deal with the non-negative members of M_ρ^* , since the general
results will follow from the usual decompositions.

LEMMA 5. <u>Given</u> $0 \le x^* \in M_\rho^*$, <u>define</u> $v(E) = x^*(\chi_E)$ <u>for</u>
$E \in \Sigma_0$. <u>Then</u> v <u>is a non-negative finitely additive set</u>
<u>function on</u> Σ_0 <u>which vanishes on μ-null sets and is bounded</u>
<u>by</u> $v(E) \le \|x^*\|\rho(\chi_E)$.

Proof. The set function v is clearly non-negative and
finitely additive. Moreover, $0 \le v(E) = x^*(\chi_E) \le \|x^*\|\rho(\chi_E)$.
In particular, if $\mu(E) = 0$, then $\rho(\chi_E) = 0$ and thus $v(E) = 0$.

QED.

Let us fix $0 \le x^* \in M_\rho^*$ and the resulting v as given
in the above lemma. Now consider a μ-simple function $h \in M_\rho$,
say $h = \sum_{i=1}^{n} \alpha_i \chi_{E_i}$ with $\{E_i\}$ a disjoint collection in Σ_0 .
Then h is also v-simple. Moreover, by the definition of
integration with respect to a finitely additive set function
[6, 2],

$$x^*h = \sum_{i=1}^{n} \alpha_i x^*(\chi_{E_i}) = \sum_{i=1}^{n} \alpha_i v(E_i) = \int h\,dv .$$

An immediate extension of the integration to arbitrary elements
of M_ρ cannot be made since Σ_0 is not a field. By working

in $(\Omega_n, \Sigma_n, \mu)$, however, the desired type of integration will be attained.

THEOREM 6. **If** $f \in M_\rho(\Omega_n, \Sigma_n, \mu)$, **then** f **is integrable relative to** ν , **and** $\int_{\Omega_n} f d\nu = x^*(f)$.

Proof. In order for f to be ν-integrable [6] on Ω_n , there must exist a sequence $\{f_n\}$ of ν-simple functions such that

$$\text{i)} \quad f_n \to f \quad \text{in } \nu\text{-measure,}$$

$$\text{and} \quad \text{ii)} \quad \lim_{m,n\to\infty} \int_{\Omega_n} |f_n - f_m| d\nu = 0 .$$

Since $f \in M_\rho(\Omega_n, \Sigma_n, \mu)$ there exists $\{f_n\}$ which are μ-simple and hence ν-simple on Ω_n such that $\rho(f - f_n) \to 0$.

To see i) we set $T_n^\alpha = \{\omega \mid |f_n(\omega) - f(\omega)| > \alpha\}$ for $\alpha > 0$. Since f_n and f are measurable, $T_n^\alpha \in \Sigma$; moreover, since the functions are restricted to having support in Ω_n , $T_\alpha^- \in \Sigma_n$, for n=1,2,... . Therefore $x^*(|f_n - f|) \geq x^*(|f_n - f| \chi_{T_n^\alpha}) \geq x^*(\alpha \chi_{T_n^\alpha})$

$= \alpha \nu(T_n^\alpha)$. Consequently, since $x^*(|f_n - f| \leq \|x^*\| \rho(f_n - f) \to 0$, we have that $\nu(T_n^\alpha) \to 0$ for any $\alpha > 0$. Hence $f_n \to f$ in ν measure. To see ii), we recall that if h is simple with support in Ω_n then $\int_{\Omega_n} h d\nu = x^*(h)$. Thus

$$\int_{\Omega_n} |f_n - f_m| d\nu = x^*(|f_n - f_m|) \leq \|x^*\| \rho(f_n - f_m) \to 0 .$$

Consequently, $f \in M_\rho(\Omega_n, \Sigma_n, \mu)$ is ν-integrable and (by definition) $\int_{\Omega_n} f d\nu = \lim_n \int_{\Omega_n} f_n d\nu$. It follows that

$$\left| x^* f - \int_{\Omega_n} f d\nu \right| \leq \left| x^* f - x^* f_n \right| + \left| x^* f_n - \int_{\Omega_n} f_n d\nu \right| + \left| \int_{\Omega_n} f_n d\nu - \int_{\Omega_n} f d\nu \right| \to 0$$

as $n \to \infty$.   QED.

LEMMA 7. ν <u>is of finite variation on any set</u> $E \in \Sigma_0$, <u>and the variation is bounded on</u> Ω_n .

Proof. We denote by $\upsilon(\nu, E)$ the variation of ν on E . Then for $E \in \Sigma_0$, $\upsilon(\nu, E) = \sup \left\{ \sum_{i=1}^{n} |\nu(E_i)| \,\Big|\, E_i \subseteq E \, , \, \{E_i\} \text{ disjoint} \right\}$

$$= \sup \sum_{i=1}^{n} |x^* \chi_{E_i}|$$

$$= \sup_i |x^* \chi_{\cup E_i}|$$

$$\leq \|x^*\| \, \rho(\chi_E) \quad < \infty .$$

Moreover, if $E \in \Sigma_n$, then $\upsilon(\nu, E) \leq \|x^*\| \, \rho(\chi_{\Omega_n})$.   QED.

Thus far the development has been for a fixed but arbitrary member Ω_n of a ρ-admissible sequence. Since every member of Σ_0 is an element of some admissible sequence, it has been shown that, given $0 \leq x^* \in M_\rho^*$, the resulting ν satisfies $x^* f = \int_\Omega f d\nu$ for all $f \in M_\rho$ whose support is in Σ_0 . We will now extend this to all of M_ρ .

DEFINITION 8. Let $0 \leq x^* \in M_\rho^*$ and let ν be the set function defined as $\nu(E) = x^*(\chi_E)$, $E \in \Sigma_0$. Given $f \in M_\rho$, there exists a sequence $\{f_n\}$ in M_ρ such that f_n has support in Ω_n and $\rho(f-f_n) \to 0$, where $\{\Omega_n\}$ is a ρ-admissible sequence. Define $\int_\Omega fd\nu = \lim_n \int_{\Omega_n} f_n d\nu$.

REMARK. This integral is well defined; in particular, it is independent of the sequences chosen.

THEOREM 9. <u>For an</u> $x* \in M_\rho^*$ <u>there exists a finitely additive set function</u> ν <u>on</u> Σ_0 <u>such that</u> $x*f = \int_\Omega fd\nu$ <u>for all</u> $f \in M_\rho$. <u>Moreover,</u> ν <u>is of finite variation on any</u> $E \in \Sigma_0$.

We will now effect the reverse.

DEFINITION 10. Let $\mathcal{U} = \{\nu \mid \nu$ is real-valued additive set function on Σ_0, which vanishes on μ-null sets, and which has $\|\nu\| < \infty\}$, where $\|\nu\| = \sup \{|\int fd\nu|/\rho(f) \mid f$ simple$\}$.

THEOREM 11. $\mathcal{U}$ <u>is a linear space and</u> $\|\cdot\|$ <u>is a norm on</u> $\mathcal{U}$.

THEOREM 12. $M_\rho^* \cong \mathcal{U}$.

Proof. It is already established that given $x* \in M_\rho^*$ there exists an additive set function defined on Σ_0 by $\nu(E) = x*(\chi_E)$. Moreover, if f is simple, then $|\int fd\nu| = |x*f| \le \|x*\|\rho(f)$. Consequently $\|\nu\| = \sup\{|\int fd\nu|/\rho(f) \mid f$ simple$\} \le \|x*\|$, so that $\nu \in \mathcal{U}$.

Conversely let $\nu \in \mathcal{U}$. (As usual, it suffices to consider $\nu \ge 0$.) If f is simple, define $x*f = \int fd\nu$. Moreover $|x*f| = |\int fd\nu| \le \rho(f)\|\nu\|$. Since the simple functions

are dense in M_ρ , x^* has a unique norm-preserving extension
to all of M_ρ , designated as $x^*f = \int f d\upsilon$. Linearity of x^*
is clear and $\|x^*\| \leq \|\upsilon\|$ establishes that $x^* \in M_\rho^*$.

It is clear that the maps of the two preceding paragraphs
are linear and lattice isomorphisms which are inverse to each
other.   QED.

COROLLARY 13. $\mho$ __is complete.__

DEFINITION 14. The space $\beta_{\rho'}$ is defined as $\beta_{\rho'} = \mho \times P_{\rho'}$,
with norm $\|(\upsilon,\psi)\| = \|\upsilon\| + |\psi|(\Omega)$ and partial order $(\upsilon,\psi) \geq (0,0)$
if $\upsilon \geq 0$ and $\psi \geq 0$, where $\upsilon \in \mho$ and $\psi \in P_{\rho'}$.

THEOREM 15. __If__ L_ρ __has property (I), then__ $L_\rho^* \cong \beta_{\rho'}$.

A more explicit representation of the functionals of the
above theorem can be given. This will now be demonstrated.
Recall that the correspondence (Theorem 1) between M_ρ^* and
$(M_\rho^\perp)^{o.c.}$ is effected by taking the correspondent of $\overline{y}^*$ in
M_ρ^* to be the Hahn-Banach extension of $\overline{y}^*$ to all L_ρ . It
was noted that this is unique modulo elements of $M_\rho^\perp$. In
fact, however, this must be unique and belong to $(M_\rho^\perp)^{o.c.}$,
since if w^* is such an extension, then $\|\overline{y}^*\| = \|w^*\|$ and
$\overline{y}^*(f) = w^*(f)$ for all $f \in M_\rho$. Moreover, $w^* = y^* + z_1^*$
where $y^* \in (M_\rho^\perp)^{o.c.}$ and $z_1^* \in M_\rho^\perp$. (If y^* were not the
same element in this decomposition, then there would be more
than one member of $(M_\rho^\perp)^{o.c.}$ corresponding to $\overline{y}^*$ which
contradicts the fact that the correspondence is one-to-one.)
But Corollary 2 states that $\|w^*\| = \|y^*\| + \|z_1^*\|$, which implies

$\|z_1^*\| = 0$, since both $\|w^*\|$ and $\|y^*\|$ are equal to $\|y^*\|$.
Hence $z_1^* = 0$ and $y^* = w^*$.

This fact will now be used to define an integral of elements
of L_ρ with respect to members of $\mho$, which will give pre-
cisely the values that the correspondent of $\mho$ in $(M_\rho^\perp)^{o.c.}$
gives when applied to f . It suffices to consider the case
for positive f and v , so let $0 \le f \in L_\rho$ and $0 \le v \in \mho$.
(Then there is a functional $\overline{y}^*$ in M_ρ^* corresponding to v
such that $\overline{y}^*(f) = \int_\Omega fdv$ and $v(E) = \overline{y}^*(\chi_E)$.) Let f_n be
the truncations of f , i.e. take $\{\Omega_n\}$ to be a fixed but
arbitrary ρ-admissible sequence, and define

$$f_n(\omega) = \begin{cases} f(\omega) & \text{if } \omega \in \Omega_n \text{ and } f(\omega) \le n \\ 0 & \text{otherwise.} \end{cases}$$

Then $f_n \uparrow f$ and each $f_n \in M_\rho$. It is standard that $\int_\Omega f_n dv$
exists (in the usual sense) and $\int_\Omega f_n dv \le \rho(f_n)\|v\|$. Moreover,
since $v \ge 0$ and $f_n \uparrow$, the sequence of the values defined by
the integrals is increased and bounded above by $\rho(f)\|v\|$.
Define $\int fdv = \lim_n \int f_n dv$. For general f and v , define
$\int fdv$ by linearity on the decompositions of f and v into
positive and negative parts. A little computation shows that
this is linear in f . Thus defining $x_\tau^*(f) = \int fdv$ gives a
linear functional on L_ρ which has norm less than or equal to
$\|v\|$. But, if $f \in M_\rho$, then $f_n \to f$ in norm, also, and thus
$\int_\Omega fdv$ exists in the usual sense; i.e. $x_\tau^*(f) = \int_\Omega fdv$, for $f \in M_\rho$,

where this integral is the usual one. Thus x_τ^* agrees on M_ρ
with the correspondent $\overline{y}^*$ for v in M_ρ^* so that $\|x_\tau^*\| = \|v\|$.
But this means that x_τ^* is a Hahn-Banach extension for $\overline{y}^*$,
hence is the unique extension and is in $(M_\rho^\perp)^{\text{o.c.}}$.

The preceding can be summarized as:

COROLLARY 16. <u>In the correspondence of Theorem 15, one</u>
<u>has</u> $x*f = \int f dv + I_\psi(f)$, <u>where</u> ψ <u>corresponds to</u> $z*$, <u>the</u>
<u>component of</u> $x*$ <u>in</u> $M_\rho^\perp$, <u>and</u> v <u>corresponds to</u> $y*$, <u>the</u>
<u>component of</u> $x*$ <u>in</u> $(M_\rho^\perp)^{\text{o.c.}}$.

Before concluding the paper, a common phenomenon in representations will be examined; viz, which elements of $\mho$ are
countably additive?

DEFINITION 17. Define $\mathcal{C} = \{v \in \mho : v \text{ is countably addi-}$
tive on $\Sigma_0\}$.

It is clear that $\mathcal{C}$ is a subspace of $\mho$; that it is closed
follows from the Nikodym corollary to the Vitali-Hahn-Saks
Theorem [6].

THEOREM 18. $\mathcal{C} \cong L_{\rho'}$.

Proof. If $v \in \mathcal{C}$, then since v is countably additive
and vanishes on μ-null sets, v is μ-continuous [6]. Then by
the Radon-Nikodym Theorem, there is a unique μ-integrable
function g such that $v(E) = \int_E g d\mu$, $E \in \Sigma_0$. Letting $x*$
be the correspondent of v in Theorem 12, we have $x*f = \int f g d\mu$
for $f \in M_\rho$. Moreover,

$$\rho'(g) = \sup\left\{ \left| \int fg d\mu \right| / \rho(f) \,\Big|\, f \in L_\rho \right\}$$

$$= \sup\left\{ \left| \int fg d\mu \right| / \rho(f) \,\Big|\, f \in M_\rho \right\}$$

$$= \sup\left\{ \left| \int f d\nu \right| / \rho(f) \,\Big|\, f \in M_\rho \right\}$$

$$= \|\nu\| \ .$$

Thus $g \in L_{\rho'}$ and $\rho'(g) = \|\nu\|$.

Conversely, if $g \in L_{\rho'}$, define $\nu(E) = \int_E g d\mu$ for $E \in \Sigma_0$. Then ν is countably additive. If $f \in M_\rho$, then f is ν-integrable, with $\int f d\nu = \int fg d\mu$. Finally by reversing the computation in the previous paragraph $\|\nu\| = \rho'(g)$.
Thus $\nu \in C$.   QED.

THEOREM 19. $C = U \Leftrightarrow M_\rho = L_\rho^\alpha$.

Proof. The previous theorem has established that $C \cong L_{\rho'}$. Assume $M_\rho = L_\rho^\alpha$. Since, in general, $L_\rho^\alpha \subseteq L_\rho^\pi \subseteq M_\rho$ for any admissible sequence π , in this case $L_\rho^\alpha = L_\rho^\pi = M_\rho$ for every π . Thus by I.2.11 (iii), $M_\rho^* = (L_\rho^\alpha)^* \cong L_{\rho'} \cong C$. Thus $M_\rho^* \cong C$ and consequently $U = C$.

Conversely, assume that $U = C$. Then it suffices to show that $L_\rho^\alpha \supseteq M_\rho$. By assumption $L_{\rho'} \cong C = U \cong M_\rho^*$; i.e. $L_{\rho'} \cong M_\rho^*$. Moreover M_ρ is a closed subspace of L_ρ with the property that $f \in M_\rho$ implies $\mathrm{Re}f$, $\mathrm{Im}f$, and $f\chi_E \in M_\rho$. A theorem generalizing I.2.11 (iii) in [19] gives that $M_\rho \subseteq L_\rho^\alpha$.

QED.

Bibliography

1. Ando, T., "Linear functionals on Orlicz spaces," Nieuw.
    Arch. Wisk. (3), 8(1960), 1-16.

2. Bartle, R. G., "A general bilinear vector integral,"
    Studia Math., 15(1956), 337-352.

3. Bochner, S and Phillips, R. S., "Additive set functions
    and vector lattices," Ann. Math. (2), 42(1941) 316-324.

4. Day, M. M., _Normed Linear Spaces_, Academic Press, New York,
    1962.

5. Dunford, N. and Pettis, B. J., "Linear operations on sum-
    mable functions," Trans. Amer. Math. Soc., 47(1940),
    323-392.

6. Dunford, N. and Schwartz, J. T., _Linear Operators, Part I:_
    General Theory, Interscience, New York, 1958.

7. Ellis, H. W. and Halperin, I., "Function spaces determined
    by a levelling length function," Can. J. Math., 5(1953),
    576-592.

8. Halperin, I., "Function spaces," Can. J. Math., 5(1953),
    273-288.

9. Kakutani, S., "Concrete representation of abstract (L)
    spaces and the mean ergodic theorem," Ann. of Math. (2),
    42(1941), 523-537.

10. Kakutani, S., "Concrete representations of abstract (M)
    spaces," Ann. of Math. (2), 42(1941), 994-1024.

11. Kantorovitch, L. F. and Vulich, B. Z., "Sur la représenta-
    tion des opérations linéares," Compositio Math., 5(1938),
    119-165.

12. Kelley, J. L. and Namioka, I., _Linear Topological Spaces_,
    Van Nostrand, Princeton, 1963.

13. Köthe, G., "Die Teilräume eines linearen Koordinatenraumes,"
    Math. Ann., 114(1937), 99-125.

14. Köthe, G., "Die Stufenräume, eine einfache Klasse linearer
    vollkommenen Räume," Math. Z., 51(1948), 317-345.

15. Köthe, G., "Neubegrundung der Theorie der vollkommenen
    Räume," Math. Nachi., 4(1951), 70-80.

16. Köthe, G. and Toeplitz, O., "Lineare Räume mit unendichvielen Koordinaten," J. Reine Angew. Math., 171(1934), 193-226.

17. Krasnosel' skii, M. A. and Rutickii, Ya. B., _Convex Functions and Orlicz Spaces_, (Translation). P. Noordhoff Ltd., Groningen, 1961.

18. Lorentz, G. G. and Wertheim, D. G., "Representation of linear functionals on Köthe Spaces," Can. J. Math., 5(1953), 568-575.

19. Luxemburg, W. A. J., "Banach function spaces," Ph.D. Thesis, Delft, 1955.

20. Luxemburg, W. A. J. and Zaanen, A. C.[1], "Notes on Banach function spaces," Proc. Acad. Sci. Amsterdam (Indag. Math.), Note I, A66(1963), 135-147; Note II, A66(1963), 148-153; Note III, A66(1963), 239-250; Note IV, A66(1963), 251-263; Note V, A66(1963), 496-504; Note VI, A66(1963), 655-668; Note VII, A66(1963), 669-681; Note VIII, A67(1964), 104-119; Note IX, A67(1964), 360-376; Note X, A67(1964), 493-506; Note XI, A67(1964), 507-518; Note XII, A67(1964), 519-529; Note XIII, A67(1964), 530-543; Note XIV, A68(1965), 229-248; Note XV, A68(1965), 415-446; Note XVI, A68(1965), 646-667.

21. Luxemburg, W. A. J. and Zaanen, A. C., "Some remarks on Banach function spaces," Proc. Acad. Sci. Amsterdam (Indag. Math.), 59(1956), 110-119.

22. Luxemburg, W. A. J. and Zaanen, A. C., "Compactness of integral operators in Banach function spaces," Math. Ann., 149(1963), 150-180.

23. Luxemburg, W. A. J. and Zaanen, A. C., "Some examples of normed Köthe spaces," Math. Ann., 162(1966), 337-350.

24. Morse, N. and Sacksteder, R., "Statistical Isomorphism," Ann. Math. Statistics 37(1966), 203-214.

25. Nakano, H., _Modulared Semi-Ordered Linear Spaces_, Maruzen, Tokyo, 1950.

26. Nakano, H., _Linear Lattices_, Wayne State University Press, Detroit, 1966. (This is part of [25] reprinted.)

---

[1] Notes XIV, XV, and XVI are by W. A. J. Luxemburg alone. Though all the papers of [20] are not used in the present paper, they are included here since they provide a complete list of the Notes.

27. Phillips, R. S., "On linear transformations," Trans. Amer.
     Math. Soc., 48(1940), 516-541.

28. Rao, M. M., "Linear functionals on Orlicz spaces," Nieuw.
     Arch. Wisk., 12(1964), 77-98.

29. Rao, M. M., "Linear functionals on Orlicz spaces: general
     theory," (to appear).

30. Rosenbloom, P. C., "Quelques classes de problemes extremaux,"
     Bull Sci. Math. France, 79(1951), 1-58 and 80(1952),
     183-215.

31. Schaeffer, H. H., _Linear Topological Spaces_, MacMillan,
     New York, 1965.

32. Taylor, A. E., _Introduction to Functional Analysis_, John
     Wiley and Sons, New York, 1958.

33. Uhl, J. J., "Orlicz spaces of additive set functions and
     set-martingales," Ph.D. Thesis, Carnegie Institute of
     Technology, 1966.

34. Weiss, G., "A note on Orlicz spaces," Portugaliae Math.,
     15(1956), 35-47.

35. Yosida, K. and Hewitt, E., "Finitely additive measures,"
     Trans. Amer. Math. Soc., 72(1952), 46-66.

36. Zaanen, A. C., _Linear Analysis_. P. Noordhoff, Groningen,
     and Interscience, New York, 1953.

37. Zaanen, A. C., "The Radon-Nikodym theorem I, II," Indag.
     Math., 23(1961), 157-187.